INFORMATION TECHNOLOGIES
AND
SOCIAL ORDERS

COMMUNICATION AND SOCIAL ORDER

An Aldine de Gruyter Series of Texts and Monographs

Series Editor

David R. Maines, Wayne State University

Advisory Editors

Bruce Gronbeck • Peter K. Manning • William K. Rawlins

INFORMATION TECHNOLOGIES
AND
SOCIAL ORDERS

Carl J. Couch

Edited with an Introduction by David R. Maines
and Shing-Ling Chen

ALDINE DE GRUYTER
New York

About the Author

Carl Couch (1925–1994) received his PhD. from the University of Iowa and served on the faculty of the University. Dr. Couch was President of the Society for the Study of Symbolic Interaction (1982), which he helped establish. He also served as President of the Midwest Sociological Society (1993). Dr. Couch is known as the creator of the New Iowa School of Symbolic Interaction.

About the Editors

David R. Maines is a faculty member in the Department of Sociology, Wayne State University. Dr. Maines serves as series editor of the Communication and Social Order Series for Aldine de Gruyter. He has worked to articulate an interactionist approach to the study of social organization and to the fundamental relevance of temporality and communication for sociological analysis.

Shing-Ling Chen is an Assistant Professor of Mass Communication at the University of Northern Iowa. Her research interests include the study of the impacts of electronic media on community structures and audience experiences, and the construction of social relationships in cybersociety.

ALDINE DE GRUYTER
A division of Walter de Gruyter, Inc.
200 Saw Mill River Road
Hawthorne, New York 10532

This publication is printed on acid free paper

Library of Congress Cataloging-in-Publication Data

Couch, Carl J.
    Information technologies and social orders / Carl J. Couch ;
  edited with an introduction by David R. Maines and Shing-Ling Chen.
        p.   cm. — (Communication and social order)
    Includes bibliographical references and index.
    ISBN 0-202-30515-5 (cloth : alk. paper). — ISBN 0-202-30516-3
  (paper : alk. paper)
        1. Communication—Social aspects. 2. Information technology—
  Social aspects. 3. Information society. I. Maines, David R.
  II. Chen, Shing-Ling. III. Title. IV. Series.
  HM258.C678   1996
  302—dc20   96-14824
                                                              CIP

Manufactured in the United States of America

10   9   8   7   6   5   4   3   2   1

We did as you asked, Carl
Rest well

Contents

Information and Action: An Introduction to Carl Couch's Analysis of Information Technologies

DAVID R. MAINES and SHING-LING CHEN

An information technology contributed directly to the transformation of the social structures of Middle Eastern societies several thousand years ago, and it appears that this may well be happening yet again. Indeed, as Rogers (1986:244–45) describes, the Iranian revolution of 1979 was significantly a conflict between big and little media. The Shah, of course, had sought to modernize Iran, in the course of which he had massively invested in cutting-edge information technologies—broadcasting equipment, microwave relays, and satellite hookups, in addition to military equipment and nuclear energy. By virtue of electronic mass media, with promodernization and proshah messages reaching 100 percent of Iranians by radio and 70 percent by television, the ideological and religious fractures between the elite and poor majority widened. That fracture, in its consequence, solidified a market for the exiled Ayatollah Khomeini, who had moved to France in the early 1960s to take advantage of the direct-dial telephone system that had been installed in Iran. For fifteen years prior to the revolutionary takeover, Khomeini used that system and sent daily messages to his followers by audiotapes, which then were copied and played on tape recorders and also transcribed and distributed in mimeograph and photocopied form. It was the use of the tape recorder, in effect, that kept the antishah movement alive, kept an exiled charismatic leader in place as a leader, and helped to mobilize mass antigovernment strikes and demonstrations that eventually led to the downfall of the Shah and his government.

Now, fifteen years after the Khomeini takeover, it is the Internet that may significantly contribute to the undoing of the theocracy that the revolution initiated (Bogert 1995). Outside Israel, Iran has the fastest-growing comprehensive on-line computer system in the Middle East. Its only current direct link is through Vienna, ironically funded by the Iranian government, but several private computer companies are ready to add additional hookups. In recognition that they are losing control over information, the mullahs who control the government have instituted censorship processes, such as posting messages that the public mails (including on-line users) must conform to Islamic laws and traditions, and a division of the telecommunications ministry even has gone so far as cutting the lines of a private firm. Yet, as recent events teach us, it will be very difficult for officials to control information flow in Iran, especially when hackers, who by definition are outside conventional institutional structures, get involved. Accordingly, with computerized information technology already in place and growing, the theocratic base of Iranian political control is very likely to eventually erode.

Information technologies are not only part and parcel of political order- ings and reorderings, but are obviously central to other large-scale activ- ities such as intergroup warfare and conquest. The military success of the Mongols, as a case in point, depended not only on tactical superiority, mobility, and weaponry, but upon an effective communications network (Fairbank and Reischauer 1989:166; Grousset 1970:224–26). In addition to using the customary procedures of spies and advance envoys and scouts, Mongol post stations were established across Central Asia. These stations were used as routes of trade and protection against banditry, but signifi- cantly they functioned as information conduits for governmental and military intelligence. Using a relay of horses, Mongol couriers could cover two hundred miles in a day. Depending upon the distances involved, military information could thus be conveyed in a matter of days, or, at the outside, a couple of weeks. This information system was a remarkable feat, considering that the thirteenth-century Mongol empire was the sin- gle largest political territory in human history, and covered all of China, Central Asia, the Middle East, most of what is now Russia, and parts of Eastern Europe. And, as scholars agree, the control of information was central to Mongol military and political control over those vast territories.

While oral and written information was used by the Mongols to plan and execute military strategies, the electronic media are used today for the same purposes (Altheide 1995:179–212). The Gulf War, in fact, was the first satellite war, with both military and television signals traveling through the airwaves. But because of that very technology, the invasion phase of war, either in the form of ground troops or air attack, may well be in the incipient phase of obsolescence. "Infowar" (Waller 1995) now is completely possible, meaning that the functioning of another nation can be destroyed simply by controlling its information processing. Already used on a limited scale by the United States during the Gulf War, electron- ic technology now exists with the capability of (1) controlling a target nation's air traffic control systems, thereby rerouting its aircraft; (2) intro- ducing computer viruses into phone systems, causing massive failure; (3)

altering bank computer systems, resulting in financial chaos; (4) jamming government television systems and inserting visual and textual information that shows a leader making unpopular announcements; and (5) issuing orders to enemy soldiers that render them ineffective. If wars are ultimately won or lost not by body counts per se but by rendering economic, military, and political structures unable to act in a concerted manner, then it is altogether reasonable to envision future wars fought primarily by computers rather than with guns.

Carl Couch (1925–1994) would have been completely at ease in a discussion of these issues regarding the Iranian revolution and intergroup warfare. And he would have taken that discussion far beyond our cursory and oversimplified descriptions. For, indeed, as the last chapter of his book makes clear, Carl understood fully the implications of the term *texere,* meaning to construct or weave, which is the Latin root of the word *technology.*

Most social scientists who have investigated the social dimensions and consequences of technology have used the term to refer to some sort of production process. Industrial and agricultural production come to mind, which certainly are not inappropriate uses of the term, but they exemplify the narrower sense of its meaning. Rather, technologies are creative and sustaining processes that must find their way into human relationships and societal configurations if they are to have the effects that are so often attributed to them [see Blumer (1990) on this point]. Technologies are designs for action, but that action always takes place in a social context, and thus some other nontechnological action is necessary for technologies to do the work asked of them. Industrialization, for instance, requires a prior-existing labor force and at least a minimal market, and industrial expansion usually requires a system of credit, which itself presumes the existence of a monetary system. The point, of course, is not difficult to grasp once it is thought through, but many analysts seem to have missed that elementary idea and have given in to the seductive simplicity of technological determinism.

In *Information Technologies and Social Orders,* the nearly finished but still untitled study he left at the time of his death, Carl Couch avoids the deterministic explanations that are found in many analyses of technology and even in some analyses of language. He understands that information technologies are designs for action that extend the human senses, especially seeing and hearing, and thereby possess certain transformative capabilities. Drawing from Marshall McLuhan but probably more so from Harold Innis, he argues that different information technologies contain internal logics of their own that, in some sense, prefigure their use. These are what Innis called "biases," but are just as well thought of as sets of probabilities that are inherent in any design, blueprint, consensual plan, or tool. We can anthropomorphize the matter and say that texts *want* to be read, the spoken word *wants* to be heard, the cinema *wants* to be seen, and so on down the line with any type of technology. Expressing the point this way, of course, it is literally untrue, but Carl would agree with us that the biases that are inherent in information technologies operate more or less in these ways. Technologies, that is, call out some responses and not others, although not in a deterministic way, since sentences and cinema in and of themselves do not determine their interpretations.

If Carl was any kind of determinist, he was a social determinist, and a soft one at that, and we think that he wants us to understand that information technologies are extrapolations of the more fundamental fact of human symbolization. He does not use the term *symbol* in any esoteric sense, and certainly not a metaphysical one, but only to refer to processes of representation. Carl drew heavily from the American pragmatists, especially George H. Mead and then John Dewey, and so he had a thoroughly social theory of symbolic communication. As he makes clear, symbols are interpretations—representations of things that emerge in a social process of what Anthony Giddens (but not Carl) would call "structuration." Carl would have spoken instead of a social act, which he regarded as a social process of people doing things together. Symbols emerge from and are nested in the social relationships of human beings, and those relationships, accordingly, are the only fruitful points of departure if one wants to comprehend human symbolization (Saxton 1995).

Information technologies—speaking, quantification, writing, photographs, recordings, printing, telegraph, radio, photocopying, telephones, cinema, television, computers, Internet, satellite dishes, and so on—are designs for communicative action that, when used, transform the people and situations caught up in that action. They are symbolizations of symbols, or second-order symbols, that are formatted and reified in ways that potentiate different forms of association. Couch uses the heuristic distinction between evocative and referential symbols (we call these heuristics, since Carl and everyone else knows that there are other kinds of symbols as well) as a way to frame an analysis of these social processes. As he makes clear in a wide variety of substantive cases, the emergence of a particular information technology occurs in societal contexts that contain both the potential for that technology's adoption and sources of resistance to it. Orality, for example, privileges evocative symbols and fosters certain forms of association, whereas literacy privileges referential symbols and fosters other forms. The transformation of societies from oral-based information technologies to ones based on literacy thus was discordant (Snell 1982). There was resistance to literacy and, in a real sense, a prejudice against it. Those transformations, however messy they might have been, entailed emergent forms, that is, social facts that were not there before or that existed only in translucent or incipient configurations such as bureaucratic relationships.

Furthermore, context has everything to do with processes of emergence, even when dealing with the same information technology. Print technologies, which pertain to what Couch calls referential symbols, were developed both in China and Europe. The consequences of those developments, however, were profoundly different. Print technology in China arose in the context of state bureaucracies, and it was used almost solely for bureaucratic (e.g., record-keeping) purposes. It therefore did not diffuse throughout the Chinese population but instead remained as an insular capacity of only the few. Later in Europe, on the other hand, print technologies developed in a market context. Private, commercial print enterprises grew very rapidly in sixteenth-century Europe, with the consequence that the relations among religious, state, and scientific authorities were altered. In addition, as a number of scholars have argued, the printing press effectively gave rise to a new phenomenon: the public, or a form of association that transcended specific localities and provided a common ground for strangers to deal with one another (Couch 1984). The emergence of publics, however, which was a long and uneven process, did not reach its fuller potential until the development of the newspaper and especially the penny press in the early 1800s. State authorities, as Couch describes, were cautious if not openly hostile to the broader base of information flow that came with the development of the newspaper. But the fate of news and its dissemination to publics varied according to whether the political order was legislature-based or monarchy-based. Again, issues of context and the sets of relations within which an information technology develops are critical to an adequate understanding of the technology itself.

We see in Couch's analysis a consistent theory of human association that is applied to a dazzling array of historical and cultural situations. His theoretical commitments demand of social scientists something that, while ontologically correct, is difficult to accomplish, namely, to use the social act as the irreducible level of analysis. Whether consisting of relations between two individuals or two state structures, it is what those two acting units do together in a social process that is important. In that act, or what Anselm Strauss (1993) has recently called "ordering processes," one finds information technologies that are the very stuff of the forms of association that make up society. This, we think, is what Carl wanted all of us to understand better. As far as he was concerned, the fate of credible scholarship hinges on that understanding, and even if one pays only scant attention to his analysis, it is just as clear that he felt that the fate of humanity hinges on it as well.

Preface

McLuhan's cliché—the medium is the message—contains a powerful truth. But that truth has been so vulgarized by so many that it has almost become a falsehood. It is a modification of that cliché—each medium offers many messages—that informs this book.

I use the term *information technologies,* instead of media, in recognition that significant modifications in information management do not always involve mass communication. The term *media* has come to be too confining for students of communication, insofar as modifications in the techniques used to preserve information are as significant as modifications in techniques used to transmit information.

Almost without exception, people who have been directly involved in the development and dissemination of new information technologies have claimed they are a boom to humanity, while those most embedded in traditional technologies have usually had their doubts about the benefits to be derived from new technologies. Most information technologies, however, have in fact expanded human experiences and enlarged human arenas. Reciprocally, information technologies have been used as instruments of exploitation. For millennia, pictographic writing was a powerful device used by the elite to exploit the rank-and-file citizens of totalitarian states. Yet the development of most information technologies, including pictographic writing, have provided human beings with new opportunities to preserve and share information and have thereby enriched human life.

The development of each information technology has contributed to transformations of the human condition. Each information technology has enhanced the memories of both individuals and collectivities, and has allowed for human beings to undertake activities not previously possible. The experiences of most citizens of contemporary civilizations are much more exciting, complex, and some might even say satisfying than those of nonliterate nomadic bands.

The discussion of the development of information technologies offered in this book is presented more or less as a chronology. The emphasis is not on the chronology per se, however, but on epochs associated with each technology. The epochs are not cleanly delineated: they fade into one another. Pictographic writing is an ancient technology, for example, that has been replaced in much of the world, but it continues to be used in modern China. And numeric concepts formulated thousands of years ago continue to infuse the communicative processes of all modern civilizations.

The specification of epochs, of course, is done with the benefit of hindsight. A few information technologies had a rather pervasive immediate impact, only to fade from the scene with few substantial consequences. A recent example of that is optical telegraphy. When the first optical telegraph network was unveiled it generated more fanfare than did the first radio broadcast, or for that manner, the first television broadcast. But visual telegraphy went the way of the dinosaurs whereas radio and television broadcasts flood the homes of nearly everyone with sounds and sights.

In recent years, several communication theorists have offered a framework for the analysis of information technologies that proceed as if information technologies were developed and used in a social vacuum. All technologies, however, both material and informational, were developed within particular social contexts and in turn influenced those contexts. An adequate theory of information technologies must therefore attend to the social context, the technologies themselves, and the interface between the technologies and their social contexts. That dictum has been the guiding principle in my analysis.

Acknowledgments

This draft has benefited from the comments and suggestions of many others. In addition, several extended discussions of the issues have contributed to the formulation offered. Those who have made important contributions include Rose Jensen, Dave Deikema, Joe Bishop, Douglas Adams, Patrick Czupik, Richard De Laurell, Ana Garner, Erling Holtsmark, Michael Katovich, Shirley Serini, Charles Abbott, Donna Flayhan, Bill Wright, Jim Stoner, Susan Baumann, Steven Couch, Stanley Saxton, Mari Molseed, David Maines, Lisa Landgraf, Barb Willard, Melissa Wood, Ralph Beliveau, Cloy Tobola, Jim Terkeurst, Julie Friedline, Jeff Houser, and a special appreciation is extended to Shing-Ling Chen. Several of the ideas, especially those in the chapters on recording and broadcasting, were developed in our many conversations about information technologies and their consequences. I have stolen some of the better ideas in those chapters from her. But that is okay, as she has stolen ideas from me. I have not kept track of who owes whom, but I suspect I owe her at least a few ideas.

CHAPTER

1

Introduction

If the continuation of human life is contingent on human beings preserving and sharing symbolic information, as clearly is the case, then changes in how people preserve and share information must have profound consequences. Despite that, only a few social theorists have analyzed the interconnectedness of communication, information technologies, and social structures. The neglect of these interconnections must be rectified if a comprehensive theory of social life is to be formulated.

Human beings not only communicate with one another, they also construct social relationships. Social relationships emerge from and inform interaction. Social relationships are relatively constant and they provide a degree of transituational consistency to human action, but they are not static. They only exist in process. They are used by people to structure actions with and with respect to one another.

The social relationships that people can construct reflect the information technologies they use and how those technologies are used in turn reflects social relationships, especially those relationships that constitute economic, religious, and state structures. Modifications in how information is shared and preserved both reflect and influence social structures.

All communication has a processual dimension, but not all information is dynamic. The written word endures in static form, but reading is a dynamic activity. Much communication, even communication based on dynamic information such as speech, contributes to the consistency of social relationships. When friends reconvene, they commonly affirm their friendship by speaking to one another. Those who have mastered a written language may affirm their friendship by writing to one another. When people affirm relationships via writing, the information transmitted is, in one sense, more constant than the relationships themselves. Love letters often endure long after romantic relationships have been terminated.

Most communication, both that which employs a technology and that which does not, is contextualized by social relationships. An exchange of letters by a student and teacher elicits different experiences and consequences than an exchange of letter by lovers even when the content of the letters is the same.

Accordingly, an adequate theory of human conduct must take as its minimal units of analyses the communicative transactions, social relationships, and processes of communication. Of course, at any point of an analysis attention may be focused primarily on communicative transactions, on social relationships, or procedures of communicating. But the analysis must be informed by a conceptual framework that recognizes that social life is multidimensional. Reciprocally, the development of a comprehensive theory of information technologies requires that attention be given to social relationships as well as the technologies used to preserve and transmit information.

COMMUNICATION PROCESSES

Symbolic information emerges from social encounters wherein people align their actions with one another as they confront and are confronted by an environment. The minimal conditions necessary for the development of the ability to communicate are two interacting human beings impinging on and being impinged on by an environment. Solitary individuals cannot survive infancy, and even if they did they would not develop the full ability to communicate.

Although the ability to communicate emerges from social interaction, each generation does not create a pool of symbolic information ex nihilo. Instead, each generation preserves some of the information that it and preceding generations have created and transmits it to following generations. Early socialization requires interactive encounters between newcomers and older persons wherein newcomers acquire the ability to communicate from older persons. Socialization is not always successful and when it is successful it does not produce automatons (Lind 1988:174).

After individuals acquire information, they become capable of symbolically organizing their personal actions and generating additional information. The ability of individuals to generate information rests on a social foundation but the generation of information is not limited to social encounters. On occasion individuals generate novel information by recombining information acquired from the older generation with information generated in transactions with their environment. The novel symbolic information sometimes is transmitted to others. More often novel information exists but a short time and vanishes without a trace.

Prior to acquiring command of symbols, individuals are capable of acting *with respect to* one another, but it is only as they acquire command of symbols that they can share experiences and act *with* one another. Presymbolic children may have experiences in common with others, but mutual consciousness emerges with acquisition of symbols. Consciousness is a social product. Shared experience and symbolically informed action are necessary prerequisites of communication, and communication, in turn, is a prerequisite of sustained coordinated action.

Most theories of communication give priority to referential symbols and slight evocative ones. Referential symbols designate objects, events, qualities, and sequences. It is by using referential symbols that people establish coorientations and coordinate their actions. Referential symbols are a prerequisite for the production of cooperative acts, but humans are sentient as well as cognitive beings. They feel as well as think, and share affect as well as thought. Evocative symbols allow people to emote in unison and with respect to one another.

Both evocative and referential symbols rest on a foundation of significant gestures. Significant gestures originally emerge in transactions wherein others assign meaning to the expressive gestures of infants. The transformation of expressive gestures into significant gestures includes three stages: (1) an infant responding to sensations (2) a caregiver noting the infant's response, and (3) the caregiver acting toward the infant in a way that changes the infant's sensations. If a number of such sequences linked to a particular expressive gesture, for example, crying, are repetitively produced, then expressive responses can be transformed into significant gestures. When the infant links its action, crying, to the subsequent action of another, it acquires the ability to call for another to act. An infant has command of a significant gesture when the infant acts—cries, then notes if another acts, and if the other does not act cries again. It requires interaction between two, an infant and another, to establish a significant gesture. The sequence, of course, is not always completed. Sometimes the other fails to hear or see the gesture; other times the other ignores the gesture. Individuals only have command of a significant gesture in conjunction with at least one other.

After significant gestures are established, the child and another are acting *with respect to* one another, but they are not sharing affect nor do they establish coorientation. They are interacting, but not communicating. Children become capable of communicating by acquiring command of significant symbols. Significant symbols allow people to be aware that they are sharing sensations and/or coorientations. After children have acquired the ability to emote and orient *with* one another toward a shared focus, they continue to emote and act *with respect to* one another. Emoting and acting with another do not exclude emoting and acting with respect to one another. We can be simultaneously depressed with and angry at the same person.

The sharing of affectivity is a more primitive social activity than the sharing of orientation. Most children acquire the ability to share sensations with another before they master the ability to establish mutual awareness of attending to the same event or object. The sharing of affect does not require coorientation. A child and another may share fearfulness without establishing mutual awareness that they are attending to the same event or object.

Affectivity, of course, is present when significant gestures are created, but at that level the child does not share affect with another. The child moves toward sharing affective experiences by expressing distress or joy with the caregiver displaying expressions of distress or joy as the child does. Repetitive production of similar expressions can culminate in children becoming aware that they and another share feelings. In the beginning the degree of sharing is minimal, but it becomes greater with social maturation.

Children acquire the ability to communicate via evocative symbols as they acquire the awareness that their displays of affect can elicit affect in others. When children acquire command of evocative symbols they may smile to elicit pleasure in another or cry to elicit stress in another. After the child understands that he shares affect with another, he can intentionally produce evocative symbols to elicit a particular affect from another. The child, that is, has the ability to act purposefully.

The infant begins to enter the realm of referential symbols when the infant and another establish mutual awareness that the two of them are attending to a common event or object and link acts to the common event or object. The establishment of referential symbols requires simultaneous attentiveness by the child and another to the same event or object and mutual awareness that both are attending to the same event or object. When that condition is established, the infant and another have a shared focus. If a shared focus cannot be established, referential symbols cannot be established.

It is only after the child has acquired awareness that he and another are simultaneously attending to a shared focus that he can produce acts that call for another to attend to a specific event or object. When a child intentionally produces an act to orient another to a specific event or object, he has entered the realm of referential communication. Referential communication establishes shared consciousness in relation to an environment.

Referential symbols often are established via vocalizations, but they are not limited to vocalizations. Pointing, for example, can establish coorientation. Vocal referential symbols are the foundation of the universe of discourse; they are vocal acts that allow people to share experiences about and act toward events and objects that are not immediately present.

In nearly all social encounters people simultaneously establish coorientations and display and elicit affect. Referential symbols allow us to document our experiences, classify them, specify relations between our experiences, and sequence the future and the past. Evocative symbols allow us to share emotions. Both evocative and referential symbols are based on shared experiences and in turn facilitate the sharing of experiences. If people are to communicate the actions or products of human experience, one person must elicit, to some degree, shared affect and / or coorientation with another.

While obviously these two types of symbols are not mutually exclusive, some communication is highly evocative and contains few referential symbols, whereas other communication consists almost entirely of referential symbols. When lovers reunite and greet each other with squeals of delight, hugs, and kisses, they primarily communicate in the universe of touch via evocative symbols. Then as they speak to each other they primarily communicate in the universe of discourse via referential symbols.

The unreflective shriek of pain is an expressive gesture; it is not an evocative symbol. But when the shriek of pain is offered with the intention of informing others of one's affective experiences, it is an evocative symbol. Shrieks may be solely an expressive gesture, both an expressive gesture and an evocative symbol, or solely an evocative symbol. The shriek of an actor performing in a play is an evocative symbol.

When one person tells another of the impending arrival of a third person who is going to physically attack them, both may become fearful. They share both the anticipation of the third person's arrival, which is based on referential communication, and the emotion of dreadfulness, which is based on evocative communication. On other occasions, shared anticipations are achieved, but disparate emotions are elicited. When a young child excitedly shouts "Mommy! Mommy!" to her father on detecting her approaching mother, she may elicit both a shared anticipation with her father toward her mother and shared emotions. But in such an instance shared anticipations may be achieved while discordant emotions are elicited. If the father has failed to do the laundry as promised, he may be distraught while the child is delighted.

When people cannot establish coorientation they may share emotions, act toward one another, and act with respect to one another, but they cannot act *with* one another. Acting with another requires establishing a shared focus, a social objective, and the sequencing of actions. For example, if two people are to hunt together, they must be able to communicate via referential symbols. At the minimum they must establish what they are going to hunt, where they are going to hunt, how they are going to align their actions with one another, and what actions they are going to take toward the prey. They can only do that if they communicate via referential symbols.

Through the production of evocative symbols people inform and misinform one another of their emotions; through the production of referential symbols they establish coorientations toward objects and events and inform and misinform one another about events, objects, and sequences. It is through evocative communication that human beings become embedded with one another and through referential communication that they establish shared foci, social objectives, and program the future. Evocative communication is the cornerstone of social solidarity; referential communication is the cornerstone of coordinated action.

The acquisition of the ability to transmit symbolic information to another—to communicate—is accompanied by the acquisition of the ability to retain symbolic information—to remember. We cannot remember any of our experiences, even intensely emotional ones, that we had prior to acquiring command of referential symbols. Our memories are composed of both evocative and referential symbols, but only those sensations that we can designate with referential symbols can be recalled.

After acquiring command of referential symbols and memories people can share personal experiences with others. For example, after a child has acquired a certain degree of sophistication with referential symbols she can inform her father of some of the experiences she had when she visited her grandparents and she and her father can share feelings as she recounts her experiences. In transactions of this sort people use referential symbols to specify emotional experiences and share affect.

Memory attenuates with time. Even our short-term memory is limited. Very few of us can accurately recall, on the bases of unaided memory, a dozen items on the shopping list we forget when we left home. If our spouse is with us perhaps we can reconstruct the forgotten list with a fair degree of accuracy. Only a very small amount of the information we are exposed to is retained with precision. For example, the precise retention of the contents of this page of print is almost impossible by unaided memory. And so it is flawed human memory that gave rise to many information technologies that were developed to increase the amount and precision of information retained.

As people extend their pasts, they increase the amount of information they remember. Some of the information retained is idiosyncratic, some is held in common, and some is shared. All of us have unique experiences, but most of our experiences are held in common or shared with others. Common and shared experiences provide the foundation of social order. Strangers, both of whom have learned to speak the same language, have only common memories. But those common memories allow strangers to share affect and to coordinate their actions. Shared pasts are acquired by people constructing an interactional history, that is, by doing things with one another, and both common and shared pasts provide collective memories. Whenever people with a collective memory establish copresence and identify one another, they use their collective memories to become socially situated.

SOCIAL RELATIONSHIPS

Collective memories are thus the foundation of social relationships. Those without memories cannot be viable members of social relationships and social structures. Those who suffer a loss of memory, such as victims of Alzheimer's disease, may be objects of attention and concern and encompassed within a social structure, but they cannot use social relationships to coordinate their actions with others.

Relationships based on shared memories emerge from interaction in which each is aware of the other as a distinct identity and both are capable of recalling that they acted with or without respect to one another. Friends activate their relationship whenever they convene and recognize one another. Relationships based on common memories are activated when people categorically identify one another and use information held in common to become socially situated. Relationships based on common memories require a common past and mutual acknowledgment of congruent categorical identities. Prototypical relationships based on common memories are those activated by two strangers at a market when they acknowledge one another and establish the congruent categorical identities of buyer and seller.

Parental relationships are a part of the social structure of all societies. Each infant must be encased by at least one parental relationship if she is to survive. Parents and others have consciousness of whether or not a specific child is their responsibility to protect, nurture, and socialize. The infant, of course, does not have consciousness of a relationship with a specific other. But as a child and a caregiver construct a shared past and the child acquires command of referential symbols, they create mutual consciousness of their relationship. After a child and parent have constructed a collective memory, they can use it to structure their encounters as soon as they establish copresence and recognize one another.

If a substantial percentage of the newborn members of a society are not encased in one or more parental relationships, a society will not endure. Parental relationships enhance but do not assure the continuation of societies. The superordinate members of parental relationships have the responsibilities of assuring the survival of subordinates and transforming the subordinates into competent members of their society. The members of parental relationships relate asymmetrically (Couch 1989:119–31), but as each parental relationship endures it tends to move toward symmetry. The interaction of mothers and infant daughters is reciprocal but very asymmetrical. As daughters mature, the interaction between them and their mothers usually becomes more symmetrical.

Solidary relationships are also panhuman. A solidary relationship is present when two or more people have a mutually acknowledged collective memory infused with affiliative emotions, project a shared distal future, and assign themselves a collective identity (Sehested 1975). Some solidary relationships such as those between lifelong friends are robust; others such as those between two acquaintances who just discovered they have an interest in common are incipient. Solidary relationships vary in extensiveness from those between two friends to nationalism.

Members of nomadic societies constructed multilayered solidary relationships as they migrated from place to place and sought the necessities of life. When they were successful in wresting the necessities of life from their environment they developed robust solidary relationships and became embedded with one another. When they were unsuccessful in their quest for the necessities of life, their solidarity ebbed and members of the band began to compete and conflict with each other. Fragmentation was a common consequence (Seekman and Couch 1989:330). Solidary relationships are not etched in stone, but robust ones are relatively enduring.

It requires two people willfully relating to one another to activate a relationship, but one person—either through unwillingness or incompetence—can prevent its activation. For example, when a child calls for her parent to help her she attempts to activate their parental relationship. If the parent refuses to acknowledge the call, the relationship, for the moment, is held in abeyance.

Fluid solidary relationships pervade the social structures of most nomadic societies. Indeed, according to Peter Wilson, kinship is present but "it is by no means the central, master organizing principle of social life" (1988:24). Rather, most relationships are based on affection and friendliness. Within such societies, solidary relationships between two people are more significant than whether the people are sisters, and commitments to one another tend to be minimal and frequently violated. The social relationships of nomadic societies are largely based on proximity, opportunity, and mutual attraction. That is, they are situational. Many social relationships in complex societies are also the consequence of proximity and opportunity, but in complex societies people more often use relationships based on categorical identities and common memories to contextualize their social encounter than is the case in nomadic bands.

There is a series of incremental graduations of the complexity of social structures from simple nomadic societies, such as the Bushmen, to the social structures that link together nations. Buyer / seller, priest / parishioner, and ruler / ruled relationships exist only in incipient form, if they exist at all, in nomadic societies. The development of those relationships was necessary for the emergence of civilizations. Economic, religious, and governmental relationships, however, continue to constitute significant dimensions of all modern nations. In some nations such as the United States, many social encounters are contextualized by economic relationships. In others, such as the Iranian theocracy, religious relationships infuse the life of citizens. In still others, such as contemporary China, governmental relationships inform much of the activity of citizens. Most contemporary adults use all three forms of relationships at various times to contextualize social encounters. Some people organize their lives largely in terms of economic relationships. For example, stockbrokers and merchants may frame most of their social encounters with mercantile concerns. Others organize their lives largely in terms of affairs of the temple, for example, devout parishioners who may relate the bulk of their activities to their religion. Still others organize their lives in terms of affairs of the state. For example, professional politicians and state officials structure much of their activity by activating categorical identities linked to state structures.

The emergence of large-scale networks of economic, religious, and governmental relationships was contingent on the creation of particular sets of categorical identities. Goods and services were exchanged long before markets were established, but in nonmarket societies the exchange of goods is submerged in other social relationships such as parental and solidary relationships. As markets became established, social contacts were made to exchange goods, mercantile interests came to the fore, and economic relationships based on the categorical identities of buyer and seller became viable features of social structures. In a nonmarket society the distribution of goods reflects relationships that rest on a foundation of a shared past. In market-dominated societies, however, the distribution of goods takes place at markets when people relate to one another on the basis of mercantile interests and the categorical identities of buyer and seller.

The social structures of modern societies obviously are complex composites of different forms of relationships. In some, state-centered relationships are paramount and economic relationships are of secondary importance. In others, economic relationships are paramount and governmental relationships are secondary. Many revolutionaries are motivated by a desire to replace the extant dominant network of relationships with a different network of relationships. Several revolutions have been dedicated to replacing a market-centered social structure with a state-centered one. In recent years several authoritarian state structures have collapsed.

Contrary to traditional anthropological lore, there is evidence suggesting that market-centered societies preceded and provided the foundation of agricultural societies (Mellart 1975, 1978; Couch 1984). The earliest sedentary communities, which appeared about 10,000 B.C., rested on a foundation of economic relationships. Market-centered societies have a more ancient history than either temple- or palace-centered ones. Wheatley (1971) advances the argument that the establishment of ceremonial centers was the watershed for the emergence of civilization. He argues that the first step toward sedentary communities was taken when nomadic food gatherers established ceremonial centers. Future research may prove that Wheatley's thesis is correct. Temples may have a more ancient past than markets, but the current evidence indicates markets antedate ceremonial centers.

It seems that relationships in which priests and temples were the focal points became pervasive only after agricultural communities were established. Large temple-centered societies emerged in western Asia about 6,000 B.C. (Adams 1966). The emergence of markets and ceremonial centers was accompanied by the rise of a public realm followed by the creation of state structures. When state structures emerged, the state functionaries assumed control of the public realm. The earliest state structures were the palace-centered civilizations established in Mesopotamia and the Nile Valley about 3000 B.C. (Hoffman 1979). In those societies relationships focused on monarchs, and palaces provided the foundation for state structures. After palace-centered structures emerged, the temple and palace elite formed a coalition and created totalitarian state structures.

The establishment of markets, temples, and palaces did not eradicate fragmentation. But societies with social structures focused on markets, temples, and palaces are capable of accumulating greater resources than are societies that do not have them. These social relationships allowed for the rise of communities that encompassed far larger populations than those of nomadic bands. Fragmentation still occurred, however, as a consequence of the collapse of markets, disaffiliations from temples, civil wars, destruction by invaders, and natural disasters.

The development of societies that rested in part on a foundation of economic, religious, and governmental state relationships allowed for the creation of social structures that were more complex than those of nomadic bands. In the long run, and it has been a very long run indeed, nomadic bands have not been able to successfully compete with communities with social structures that included complex composites of exchange, religious, and state relationships. Much of the story of humanity for the last twelve millennia has therefore been the expansion of societies with social structures linked to markets, temples, and palaces at the expense of nomadic bands.

Social relationships focused on markets, temples, and palaces tend to generate antagonistic ideologies. Those whose lives are anchored in the market develop interests that are antagonistic to the interests of those whose lives are anchored in the state structure. Traders frequently connive to avoid state regulations, or, as I have previously discussed, "The ideology of the market justifies personal autonomy, novelty (change), and contact with foreigners. Spokespersons for the temples call for homogeneity, collectivism, and constancy. Apologists for palaces rationalize hierarchy, call for obedience, and justify the use of violence against the disenchanted" (Couch 1986:153).

On many occasions, the elite of the markets, temples, and palaces have formed coalitions to advance their special positions. When that has occurred, the rank and file often have been intensely exploited. Many times the palace elite have merged with the temple elite and the solidarity generated by priests has counteracted the alienation generated in palace-centered social encounters. Palace-centered social structures rationalized by an ideology formulated by priests were often the cornerstone of ancient empires.

As social relationships that focus on markets, temples, and palaces matured, they partially replaced relationships based on shared memories. Some have bemoaned that change. But these relationships provide people with a resource that allows them to quickly become socially situated with strangers. Each time we enter a place of business and comport ourselves as a customer and another person presents himself as a clerk, we activate an economic relationship. The activation of an economic relationship allows for a far more effective distribution of goods than is possible in societies that do not have economic relationships. Members of societies with complex social structures routinely consummate complex social encounters that members of simple nomadic societies could not.

Social relationships continually evolve and devolve. The parental relationship of father / son in Western societies is much the same as it was two hundred years ago, but it has undergone some modifications. In a similar manner economic relationships, the relationships between rulers and ruled, and those between priests and parishioners have endured in much the same form for millennia. Economic relationships underwent profound changes when money was invented. In a similar manner, the relationships between state officials and citizens undergo a profound change when a monarchy is replaced by representative democracy. The citizens of the United States do not use the social relationship of monarch / subject to contextualize their encounters with state officials, but that relationship once informed the encounters between state officials and citizens of Europe.

The emergence of complex social structures was contingent on the development of information technologies as well of market-, temple-, and state-centered relationships. The development of some relationships and some information technologies is so intertwined that they might be conceptualized as a unitary phenomenon instead of two distinct phenomena. For example, the emergence of markets and that of numeric concepts were closely intertwined. Did numbers provide the foundation for markets? Or, did markets provide the foundation for numeric concepts? In some instances it is clear what preceded what. For example, the emergence of enduring palace-centered state structures was contingent on the prior development of numeric concepts and record-keeping procedures. In other instances it is difficult to specify what preceded what with more than a modicum of certainty.

INFORMATION TECHNOLOGIES

Technologies increase the ability of human beings to process matter and information. Each technology includes a pattern of action that facilitates the accomplishment of some objective. The patterns of activity that constitute a given technology may or may not include using an artifact. The use of levers allows human beings to more effectively manipulate matter; the use of written languages allows for the more effective preservation of information. Those societies that mastered material technologies were slightly more likely to survive than those which did not. Similarly, those which developed particular information technologies were more likely to survive than those that did not.

The development of later information technologies rested on a foundation of information preserved by previously developed information technologies. The development of print was contingent on the prior development of a written language. But conversely, some information technologies suppressed the blossoming of others. Pictographic writing, for example, inhibited the flourishing of print in China. Moreover, the emergence of a given technology does not assure either its endurance across generations or its adoption by other societies. Some societies, such as the Minoan, developed a written language only to subsequently lose it. Some nonliterate societies had contact with literate societies for thousands of years and did not become literate. Nor does the presence of a given technology assure the emergence of another technology. The presence of numeric concepts did not assure the development of a written language, but numeric concepts probably were a necessary precursor for the development of written languages.

All information technologies increase the amount of information preserved or in circulation or both. Each information technology also favors the preservation and dissemination of some kinds of information and disfavors the preservation and dissemination of other kinds of information. Innis (1951) was the first scholar to systematically analyze the types of information favored and disfavored by various techniques. Many subsequent scholars of communication (Ong 1982; McLuhan 1962; Carey 1989; Eisenstein 1979a, 1979b) have extended the theoretical framework originally offered by Innis. These scholars employ concepts derived from the technologies themselves to analyze media. One example is Ong's (1982) characterization of the electronic media as secondary orality. That approach makes the technologies themselves paramount, but fails to give due recognition to dimensions of communicative processes. Such dimensions vary across types of technologies and include immediate reciprocity versus delayed reciprocity, the prevalence of evocative versus referential symbols, and static versus dynamic information. For example, written language in comparison to speech promotes delayed reciprocity, enhances referential symbols, and preserves information in static form.

Nonetheless, all information technologies are contextualized by social structures and enhance or erode the social structures that contextualize them. In China the development of printing was contextualized by a palace-centered structure, while in Europe it was contextualized by a market-centered structure. The changes in the information circulated that accompanied the development of printing in China therefore promoted cultural constancy and enhanced a palace-centered state structure. However, the changes in the information circulated in Europe as printing became common, promoted cultural change, and eroded palace-centered structures.

Some communication theorists posit information technologies as the prime movers. That stance implies that social relationships are merely a reflection of information technologies. For example, Innis asserted that the printing press destroyed feudalism. Assertions of that sort have some validity, but that formulation, like all unilateral theories of social life, fails to recognize the complex interactive nature of social phenomena.

Print provided a means for infusing social structures with large quantities of novel information. Some of the social relationships current in Europe when the printing press was developed were transformed as the printing industry matured. But, conversely, one could advance the proposition that the international trade networks through which the manufacturing of paper and printing had diffused to Europe from China created the printing press in Europe.

The more fruitful approach in assessing these matters is to recognize that information technologies and social relationships have bilateral relationships. When an information technology emerges, its use will be consequential for extant social relationships, but, reciprocally, extant social relationships impact on how information technologies are used. As Blumer (1990) has argued with respect to industrialization, there are no natural interlockings of information technologies and social structures. Television broadcasting is neither inherently a commercial enterprise nor an adjunct of state structures. However, some interlockings are more congruent than others. Pictographic writing is more congruent with totalitarian state structures than with democratic state structures.

MODE OF ANALYSIS

Point-in-time analyses of social phenomena place social analysts in much the same position as a person who would attempt to determine the critical features of a cake by analyzing an already baked cake. If one wishes to determine the critical features of a cake, it is much easier to do so by observing someone baking a cake, noting the ingredients used, and the sequential order followed in mixing the ingredients together than by analyzing a baked cake. Across-time analyses focus on the sequential order of phenomena, specifies how new elements interact with extant structures, and the consequences of such interaction. Such analyses allow for the specification of necessary factors that must be in place before a given condition can be achieved.

The more ancient technologies were developed in societies that comparatively speaking possessed simple social structures. The increases in amount of information and changes in quality of information preserved by the more ancient technologies were slight, and the transformations of social relationships that accompanied their development were accordingly incremental. In contrast, some of the more recently developed technologies vastly increased the amount of information available, and their development has been accompanied by cataclysmic transformations of social structures. The complexity and rapidity of some of the more recent transformations renders analytic statements about them suspect. Nonetheless, the effort will be made. Perhaps by framing statements about recently developed technologies with a chronology that extends back to the earliest ones we may render statements about them slightly less problematic.

It is impossible to free oneself from the modern standpoint when analyzing the past. Nor is it particularly desirable to do so. Much of the analysis offered is in terms of what was not present in the past. When an analyst uses the concept of "nonliterate," she implicitly adopts the standpoint of literacy. The interpretations and conclusions derived from the comparisons offered in this book therefore reflect the biases of a literate person. If the comparisons were made by a nonliterate person, the conclusions would be different but the question of which set of conclusions is the most viable would remain. I suspect the interpretations of a literate analyst have greater viability than those of a nonliterate.

CONCLUSIONS

The uninformed human organism is not capable of surviving. If each of us had to acquire the information necessary for survival on our own, none of us would be alive. In all societies the bulk of the information used to organize conduct has been accumulated and preserved by prior generations. Communication is the core process of all human societies. Consequently a comprehensive theory of social life must attend to how information is accumulated, preserved, and shared.

Some contemporary scholars have referred to recent developments as the coming of the information age. That characterization implies that information has become more critical in social life. That, of course, is not the case. Information has been critical for the continuation of the human species for at least tens of thousands of years. But those who characterize modern technologies as ushering in an information age have a point. The recent rapid development of electronic technologies has made us more conscious of information technologies than were most prior generations, but information technologies developed hundreds, even thousands, of years ago pervade our lives as much as do television and computers now.

The continuation of human life remains contingent on copresent encounters between people wherein they simultaneously inform one another through the use of evocative and referential symbols. Interpersonal communication still remains the cornerstone of human existence. All people first acquire information in copresent encounters. It is only after children have acquired considerable information in copresent encounters that they can use information technologies to extend their memories and communicate with others. Information technologies allow us to change and sometimes improve communication, but as yet they have not entirely replaced copresent encounters. Nor does it seem likely that they will.

McLuhan's cliché "The medium is the message," although a simplification, contains a powerful truth. The emergence of each information technology has been associated with several changes in the human condition.

The issue of how information is formatted and its consequences has been recognized since Sapir and Whorf put forth the claims that variations in linguistic structures differentially frame the organization of human experiences and that the structure of the consciousness of those who speak different languages is different. The merit of those claims is not challenged. The form of the consciousness of those who speak Japanese is different from those who speak English and one might say that they live in different worlds, but the consciousness of those who live in a world of electronic broadcasting and computers is so different from those who lived in a world devoid of electronic broadcasting and computers that one might say they live in a different galaxy.

When the past is examined there is a tendency to presume it necessarily occurred so the present might occur. That is not the case. There are no forces, needs, drives, or any other factors that assure societies will develop and / or use technologies that make their continuation more likely. Some changes occurred that made the continuation of some societies more likely; but many went under. Those societies that developed specific technologies were not fulfilling some preestablished destiny. Nonetheless, how ancient technologies were and are used has overwhelming consequences for us. And the actions we undertake with our information technologies will have consequences for future generations.

CHAPTER

2

Orality

The term *oral,* which is commonly used to distinguish spoken from written communication, rests on a dichotomy that only exists in literate societies. Prior to the development of written languages, the basic dichotomy was between vernacular speech (discourse) and poetic speech (orality). A host of oral technologies were developed long before written languages were invented. Each of these oral technologies merged metered evocative symbols with referential ones to produce poetic speech. Rhythmic vocalizations that contain no referential symbols, such as nonsense lullabies, offer only patterned evocative symbols; they do not constitute an oral technology. Oral technologies combine patterned evocative symbols with referential symbols to facilitate the preservation and transmission of information.

Some societies had only one or two procedures for producing rhythmic compositions that preserved information, while others had a multitude of procedures for rendering information into poems. The Irish developed at least thirty-two different procedures for producing rhythmic compositions. Some nonliterate societies retained only a limited amount of information poetically while others preserved vast amounts of information in that form. The Greeks of the Homeric period preserved enormous quantities of information orally. The Siriono, a nomadic group regarded by anthropologists as one of the more primitive societies (Stearman 1984), however, preserved little or no information orally, but entertained themselves with nonsense songs (Holmberg 1969:111–15).

In contrast to other ancient information technologies, such as depictions, calendars, and numbers, oral technologies did not include the use of artifacts. Therefore, we will never determine with certainty when orality was first developed or how widespread its use was in preliterate antiquity. But historical and ethnographic evidence indicates that oral technologies were an important means of preserving information in most nonliterate societies and, in fact, may well be the most ancient information technology.

AS CONTRIVED SPEECH

Several different procedures for composing poetic narratives were developed. Some techniques systematically patterned the length of sounds; others systematically patterned stressed and unstressed syllables. For example, the iambic meter arranges utterances in a pattern of a series of short syllables followed by long syllables. Oral compositions were musical, and they were sung or chanted rather than spoken. The metered evocative symbols functioned to "hold the words in a fixed order" and infused the referential content of the poems with an ecstatic quality (Havelock 1971:46). The emotions that were aroused enhanced recall of the referential content and the poems generated greater emotionality than ordinary speech.

Oral recitations often were accompanied with a musical instrument—a drum or lyre—that helped to regulate the metered evocative symbols and increase their vibrancy. Hand clapping, foot stomping, and dancing also sometimes regulated the meter and increased the intensity of the emotions aroused. The emotions aroused during oral recitations were patterned evocative symbols linked to the referential content of the poems.

In modern societies, contrived patterns of speech are still produced by songwriters and poets. They merge evocative symbols and referential ones primarily to entertain, however, rather than to preserve information. Modern songs, for the most part, are composed with the aid of the written word, whereas preliterate oral compositions obviously were not.

Oral specialists could preserve far more information than is possible with noncontrived speech. For example, "One illiterate ninety-year-old man of Crete recited a poem of over 10,000 'lines'" (Culley 1967:6). The amount of information that can be retained with an oral technology is not unlimited. The upper limit of the length of an oral composition may be about the length of the Iliad, since few oral compositions have been found that exceed its length (Sarton 1952:134).

CONTEMPORARY ORALITIES

Many studies of orality have focused on the historical content of oral compositions—the exploits of preceding generations, origin myths, and narratives about the supernatural. However, most students of orality agree that oral compositions are relatively unreliable as historical documents. Henige (1974) analyzed numerous poetic renditions of the past and compared them with other evidence, and concluded that oral accounts are often modified to reflect current social arrangements. Students of orality agree that the integrity of the past is often subordinated to the integrity of the present in oral compositions (Ong 1982). For instance, in the nineteenth century an African saga was recorded that related that the tribe had been founded by seven brothers. When the same saga was recorded sixty years later the poem contained the statement that the tribe had been founded by five brothers. The tribe was composed of seven distinctive units when the saga was first recorded; only five units remained when the saga was recorded in the twentieth century (Goody and Watt 1968:33). It appears that the account of the origin of the tribe was modified in the intervening period to reflect the disappearance of two clans. Other studies have made similar observations.

If, as seems to have been the case, orally preserved accounts of the past were modified to align their content with current conditions, that practice would not provide a pool of stable information about the past. Members of such societies would be relatively unaware of how the past differed from the present, and thus would be unlikely to have a historical consciousness that included awareness of years.

In societies where orality was the dominant means for preserving information, sometimes oral specialists offered songs they knew "by heart", and other times they improvised. Not all songs were transmitted to later generations. Many simply faded away. In many societies, multiple versions of a given song had currency, and thus there was no one "correct" composition in oral societies. When songs were offered, the audience rarely was concerned with whether or not the performance was true to the original composition. If variations were noted, they would have been assessed in terms of esthetic values rather than fidelity to the original composition. The ephemerality of oral performances also operated to inhibit critical assesses of the fidelity of oral accounts.

The studies of orality that focus on the historical content of the poems suggest that accurate and precise information cannot be retained orally. But most studies of orality have analyzed sagas, compositions that were recited to audiences and offered as entertainment. When the information offered in these poems was at variance with current social conditions, it elicited disbelief from audiences. The responses of audiences enticed performers to modify the referential content of their poems to align it with current conditions.

Other studies suggest that the information preserved in oral form can remain relatively constant across generations. Minc (1988) observed that the Eskimo groups she studied used orality in two distinctive ways: one was a secular tradition; the other a sanctified ritual tradition. She found the secular tradition preserved information that informed "behaviors relating to group survival across seasonal shortages," and the sanctified tradition provided information "relevant to crisis situations recurring on the pan generation time scale" (p. 39). The compositions that constituted the sanctified tradition were contextualized by rituals, and those who recited the songs strove to render them exactly as they had been previously rendered. In contrast, the recitations of folktales and myths "took place primarily in recreational settings" and variations were common (p. 75).

The survival of food-gathering societies "is largely dependent on a group memory of past crisis situations and of the strategies appropriate for dealing with the altered environmental conditions" (p. 40). It seems that most food-gathering societies, especially those which survived in harsh environments, transmitted information in poetic form with respect to how to deal with famine. Retention of such information, of course, does not assure the continuation of a nomadic society, but it does make it somewhat more likely. Minc's research suggests that information critical to group survival remained relatively constant, but the content of other poetic compositions was less constant.

In some more complex societies, some orally preserved information also seems to have remained relatively constant across generations. Studies of the South Sea navigators, calendric specialists in contemporary Maya communities of Central America, and community leaders of southwestern Indians demonstrate that complex and precise bodies of information were preserved with a high degree of constancy by oral specialists. These studies were not undertaken to document the power of oral technologies to preserve precise and accurate information in relatively constant form. Instead they were undertaken to demonstrate the presence of complex bodies of information about celestial phenomena in nonliterate societies.

The natives of the South Pacific sailed across vast expanses of the ocean before Europeans arrived. Only a few residents of each island had command of the information required to navigate across several hundred miles of open seas. In 1922, the Gilbert Islands had a population of about thirty thousand and less than twenty men were knowledgeable about navigation (Makemson 1941:108). The navigators jealously guarded their information, but each year a few young men were selected to be instructed in navigational lore. Their instruction was shrouded in secrecy. No others were allowed to be present when the to-be-navigators were instructed.

The instructions included experienced navigators teaching poems to students. The students learned one poem to sail from island A to island B and another poem to sail from island B to island C. "Three or more priests or teachers always participated in the oral delivery of erudition in order that two might correct a slip of memory on the part of the third" (p. 274).

Each class in navigation lasted about five months. Only those "capable of reciting the teachings without a single divergence from the original were graduated" (ibid.). Those who mastered the poems were tested at sea. The students were confident that their information was adequate, and considered the longer voyages as "tests of endurance and ability to withstand hardship rather than as especially difficult navigational exercises" (Lewis 1972:158). Each poem, therefore, was like a navigational textbook. Some navigators memorized a large number of poems; others only a few. The navigators did not share their navigational information with their communities but they freely exchanged it with one another.

Similarly, Tedlock (1982) apprenticed herself to Maya *daykeepers* to learn their astronomical and timekeeping lore, and was taught poems that specified calendric sequences. "Words are linked to day names by means of paronomasia—that is, by means of poetic sound play" (p. 107). And, "Mnemonics for each day, which often embody sonic resemblances to the day name, are central to the oral system of calendar instruction and the practice of calendrical interpretation" (p. 127). The Maya daykeepers were able to determine on what day of the week the first day of the new year would fall indefinitely into the future.

Many nonliterate agricultural societies used an oral technology in conjunction with observatories to retain calendric information. One of the chants of the Hopi consisted of twenty verses that specified which crops were to be planted when the sun reached different locations on the horizon (McCluskey 1977:176). The persons in command of the calendric information also organized the annual communal celebrations. Among the Hopi, "The priest who 'keeps the calendar' in a very real sense controls pueblo activities" (Reyman 1989:18). These specialists, like the South Sea navigators, kept their knowledge secret. "Few people, if any, know exactly what priests do in carrying out the functions of their offices" (p. 23). As the priest themselves noted, "Power talked about is power lost" (ibid.).

It seems that there were at least two different sets of oral compositions in some nonliterate societies. One set of oral compositions retained information about origins, proper modes of behavior, and human exploits. Prototypes of these compositions are the Eskimo folktales and the heroic sagas of the Greeks. These compositions were typically recited for purposes of entertainment. Many of these compositions also provided moral instructions. One Eskimo folktale of an arrogant young woman stressed "the necessity for marriage and children to provide a system of social security" (Minc 1988:88). A second set of compositions in most nonliterate societies was used to organize instrumental activities critical for the survival of the societies. Prototypes of the latter were the poems of the timekeepers of the Maya who organized a yearly cycle of festivals that informed communities when to plant crops. Compositions of the latter type were typically recited in a sanctified context and usually only a few had the opportunity to learn these poems. Whereas innovations were a source of prestige in oral renditions of sagas, they were suppressed in sanctified contexts.

The social position of those who composed and recited sagas was similar to that of modern performing artists, and people attended these performances for much the same reason that modern citizens go to musical concerts—to be entertained. These specialists had prestige conferred on them if they elicited communal approval, but they did not have a monopoly of knowledge. The oral specialists who retained information that allowed them to specify when to plant crops, how and when to hold festivals, or how to sail from one island to another were not entertainers. They did not recite their compositions to general audiences. They used orally preserved information to organize the activities of their communities. In many instances the latter specialists attempted to maintain a monopoly of knowledge. Their skills often were passed on to the younger members of a family or clan. Among the traditional Hopi, "there was a basic division between those with ceremonial office and duties (high rank) and those without such responsibilities (low rank)" (Reyman 1989:11). Those with ceremonial office had command of orally preserved information; those without ceremonial office did not.

The information offered by those who recited poems in an entertaining context was neither precise nor constant. Those specialists were judged primarily on their ability to elicit excitement. In contrast, the specialists in the preservation of instrumental information were evaluated primarily on the basis of their ability to program community activities. The information of the latter had to be reasonably precise if they were to retain their special position.

Those who preserved instrumental information often combined orality with artifactually displayed information when instructing others. For example, when the to-be-navigators of the South Seas were instructed, the relative location of the islands was demonstrated to the students with arrangements of pebbles as they memorized the poems. In a similar manner, the Hopi used the orally retained calendric information in conjunction with observations of specific horizontal locations of sun risings from a central viewing position to determine when it was time to plant crops (McCluskey 1977:197).

The preconquest Maya apparently preserved calendric information in both written and oral form. Most of the written calendric information of the preconquest Maya was destroyed by the book-burning priests of their Spanish conquerors. However, a few of those books, for example the *Dresden Codex,* have survived. Those codexes demonstrate that the preconquest Maya preserved some calendric information in writing. Tedlock's research among contemporary Maya indicates that some information about solar cycles is currently retained orally.

In many nonliterate societies orality pervaded daily life. Even the Siriono, who did not use an oral technology to preserve information, sang nonsense songs in the early morning and dancing was "always accompanied by singing" (Holmberg 1969:111). During some dances a song leader emerged who chanted phrases that others repeated after him (p. 114). But it seems the Siriono did not systematically merge evocative symbols with referential ones to preserve information. They appear not to have had an oral technology.

In other nonliterate societies, oralities of various sorts infused rituals and ceremonies. For example, among the Gayo of Sumatra mourning included oral performances of ritualistic wailing, some conflicts between subvillage units were resolved by poetic duels conducted before the assembled residents, and songs were sung to welcome visitors (Bowen 1989:27). Until recently in Iceland it was a common practice in fishing communities for people to take turns reciting sagas while they worked. During the winter months as adult males worked together to repair the boats and fishing equipment, one person would recite a saga followed by another worker reciting another saga. Many of these sagas related the heroic exploits of prior generations. In some societies the emotional unity generated by a robust oral tradition was combined with a warrior tradition and fueled predatory raids on outsiders. The Greeks and Vikings were but two societies with a robust oral tradition that glorified raiding expeditions. Oral renditions of heroic exploits enhanced internal solidarity and elicited a chauvinistic attitude toward outsiders.

The presence of orality in contemporary nonliterate societies suggests that oral technologies were used in preliterate societies to preserve information. But as oral technologies left no material tracings, the case for the presence of oral technologies in ancient preliterate societies can only be made by examining the earliest writings of ancient literate civilizations for evidence that oral technologies were used in these societies before written languages were invented.

ANCIENT ORALITIES

The civilization of the ancient Greeks rested on a foundation of information that was orally retained (Havelock 1963, 1982; McLuhan 1962; Lord 1978). The civilization of the classical Greeks emerged before writing came into widespread use in Greece. Some Greek communities began using a written poetic form about 720 B.C. (Carpenter 1933:29), but prose compositions did not appear among the Greeks until about 500 B.C. and did not come into common use until "three hundred years after the invention of the alphabet had rendered the monopoly exercised by poetry over the contrived word as theoretically obsolete" (Havelock 1971:59). The Greek cities also had civic functionaries known as "rememberers" who preserved information about civic affairs in poems (Havelock 1986:84). The Greeks "did not become literate in our sense until the last third of the fifth century before Christ" (Havelock 1982:185).

Plato, whose writings mark the end of the golden age of classical Greece, bridged the transition from orality as the dominant technology to writing as the dominant procedure for preserving information (Havelock 1971). Many Hellenists have overstressed the uniqueness of the civilization of the classical Greeks, but the fact that their civilization rested on a foundation provided by orality has not been sufficiently appreciated. The oral foundation of other ancient civilizations is also underappreciated.

The orality of civilizations more ancient than the Greeks, with the possible exception of some Semitic civilizations, was replaced by pictographic writing. This process probably spanned longer durations than the replacement of orality by phonetic writing among the Greeks. Furthermore, whereas the Greeks used writing from the beginning to preserve poems, when writing was developed in more ancient civilizations it was not originally used to preserve poems. Nonetheless there are indications in the early writings of ancient pictographic societies that their written languages were preceded by a robust oral tradition.

For example, the Sumerians invented a written language about 3000 B.C., but it was not until a few hundred years after they invented writing that they transformed some of their oral compositions into written form. The most ancient Sumerian compositions were "poetic in form, ranging in length from less than fifty lines to close to a thousand" (Kramer 1959:5). Some of these compositions described the heroic exploits of past generations. Others dealt with practical problems. One ancient Sumerian poem contained instructions for the proper use of shade trees to protect garden plants (pp. 70–75). An analysis of the tales of Sinuhe, the most ancient piece of Egyptian wisdom literature, demonstrates that the tale was originally composed in oral form (Foster 1980) and the tombs of the Old Kingdom contain verses (Simpson 1973). These observations suggest that residents of the Nile valley used orality to preserve information before they became literate.

Caesar, who invaded Britain about fifty years before the birth of Christ, wrote that among the Britons there were specialists who learned "by heart a vast number of verses. Some, in consequence, remain under teaching for as many as twenty years" (as quoted in Ellis 1987:30). Caesar also noted that many of the verses dealt with celestial phenomena. "The earliest known Celtic calendar, dated from the first century AD, is far more elaborate than the rudimentary Julian one and has a highly sophisticated five-year synchronization of lunations with the solar year" (p. 32). It seems therefore that the ancient Britons employed orality to retain calendric information, but very little of that information was transformed into writing.

The Chimu and Inca civilizations were nonliterate. The Incas had professional rememberers who retained astronomical knowledge, calendric information, state records, and accounts of the Inca's past in poetic form. We have a somewhat more complete account of the orality of the Inca than most other nonliterate civilizations, because their Spanish conquerors saw fit to translate some of the poems of the Inca into writing.

Oral technologies contributed to the emergence of large and complex social structures by providing a procedure for preserving larger amounts of information than is possible with ordinary speech. These larger amounts of information allowed some ancient societies to develop complex social systems. But it is doubtful that any civilization emerged that rested solely on a foundation of orally preserved information. For example, the residents of the Mesopotamian cities of the preliterate period preserved calendric and numeric information artifactually. The alignments of their temples provided calendric information and they used tokens to preserve quantitative information. In a similar manner, the Chimu and Inca used monuments to preserve calendric information and knots in cords to preserve quantitative information about communal resources. Oral technologies, while a necessary precursor for literacy, did not themselves spawn literacy.

When written languages were invented, they did not immediately supplant orality. Writing instruments and surfaces were scarce items in ancient civilizations. The ancient Egyptians invented papyrus which was relatively easy to manufacture and, in comparison to wood and clay, an excellent writing surface. But few ancient civilization had good writing surfaces. It was not until the invention of paper that a writing surface as functional as papyrus appeared. The presence of a practical writing surface removed some of the constraints on the retention of information in writing, but until the printing press, orality remained a viable procedure for the preservation of information in all complex societies.

FORMATTING

Oral performances were vibrant. For example, when a poet of an oral society entertained an audience by singing of the exploits of past heroes, if he was an effective performer, he and his audience were elated and agonized together as he related a tale of a hero of a preceding generation. Similarly, when the timekeepers of the Hopi staged their annual celebration to herald the arrival of the spring equinox, they led their communities in chants that affirmed communal solidarity. Many societies celebrated events such as the solstices and equinoxes each year with elaborate ceremonies that contained several oral recitations. Participants in these celebrations derived both pleasure and information.

The number of people who could participate in these ancient celebrations was restricted by the range of the human voice. However, there is evidence suggesting that tens of thousands congregated at monumental centers such as Stonehenge to take part in communal celebrations. The evocative communication that must have infused these celebrations when chants were offered created and affirmed communal embeddedness. Traces of this form of activity are manifested in many modern religious and governmental rituals, including the recitation of the Lord's prayer in Christian churches and pledges of allegiance.

The predominance of evocative symbols and the ephemerality of orality generate considerable group involvement but are not conducive to adopting an analytical standpoint toward the information that is preserved. Consequently many oral compositions, especially those of considerable length, are internally inconsistent. For example, in Homer's *Iliad* there are instances where a warrior previously killed reappears to join in a battle. A concern with truth, however, emerged among the Greeks as phonetic writing replaced orality. A concern with truthfulness requires the adoption of an analytical attitude toward information that is more likely to emerge in literate societies than in oral ones.

The development of concepts (abstract symbols) is also dependent on adapting an analytical standpoint and noting similarity in two or more sets of information. Neither ordinary speech nor oral compositions is conducive to the development of concepts. Oral accounts emphasize concrete actions and events and elicit affective responses.

When orality is combined with artifactually preserved information, such as numeric inscriptions, conceptual as well as concrete information can be preserved and transmitted. While oral technologies greatly increased the amount of concrete information preserved, they did not greatly contribute to the emergence of concepts.

The dichotomy between emotional and rational action had less currency in oral societies than in literate ones. The evocative symbols established and affirmed collective sentiments; the referential ones gave direction to the collective sentiments. Emotional and cognitive responses were not regarded as antagonistic in oral societies. The distinction between the two may not even been present in ancient preliterate societies.

SPECIALISTS

In simple nomadic societies there were no oral specialists. The Eskimos did not have oral specialists; rather, all adults were presumed capable of composing poems. Of course, some were better poets than others, but the Eskimos did not categorically differentiate themselves on the basis of some having command of more information than others. The Australian Aborigines also had several different oral traditions (Muensterberger 1974:125–220), but do not seem to have had specialists in orality.

In contrast, some individuals in more complex nonliterate societies studied for years to enhance their ability to compose and recite poems. In many societies those individuals were categorically differentiated from others. Such was the case among the South Sea Islanders, Maya, Homeric Greeks, Inca, Hopi, and ancient British. Some oral specialists of these societies composed poems that required several hours to recite. Major epics were not recited at a single setting. For example, among the Greeks two or three "books" of the *Iliad* might be recited in an evening, but its complete recitation might extend over a period of a week or more.

Not all members of societies with well developed oral technologies had the ability to compose lengthy poems. Just as some literate people can only write simple compositions, some members of societies where orality flourished were able to compose only couplets. In the typical nonliterate societies with a robust oral tradition, probably only a few could compose lengthy sagas. Those capable of composing impressive poetry were accorded prestige, and they reveled in their fame much as do modern entertainers. Some oral composers and performers polished and repolished poems and their delivery of them, and labored patiently to achieve a perfect epithet (Finnegan 1988:72).

In some nonliterate societies, there were schools where techniques of oral composition were taught. In other instances, friends congregated to help one another in their efforts to compose impressive poems. In one society, a poet first performed his composition to friends who would criticize, interject comments, applaud, or howl down the performance (Grumble as quoted in Finnegan 1988:74).

Even those with the ability to compose and recite lengthy poems could only master a limited number of lengthy poems. One Yugoslavian poet claimed to be able to recite over a hundred long poems, but other Yugo-

slavian poets claimed that it was impossible for a single person to memorize such a large number of poems. They claimed that no poet was able to
recite over thirty lengthy poems (Lord 1978). A literate person, in contrast,
can read hundreds of books. The amount of information available to
literate persons with a library of a few score books is far greater than that
available to the most skilled oral technologists.

INTERFACING WITH WRITTEN LANGUAGES

Contemporary oral specialists are singularly unimpressed with writing. All who have studied oral specialists report that these specialists do
not view writing as providing new opportunities (Lord 1978; Jensen
1980). Since they had a technology that allowed them to retain any information they were inclined to preserve, some had little interest in
mastering writing. Lord (1978) found that none of the Yugoslavian oral
specialists used writing to compose poems; and none of the literate persons bothered to acquire the ability to orally compose poems. However,
when writing spread to the natives of New Zealand, some Maori poets
carried small notebooks and jotted down texts and fine turns of phrases
(Finnegan 1988:111). However, some oral specialists opposed rendering
their oral compositions into writing. The South Sea navigators were sus-
picious of efforts to render their information into written form (Lewis
1978), and the oral specialists of ancient Britain opposed translating their
poems into writing.

Most specialists were proud of their orality and regarded it as superior
to written languages. When one South Pacific specialist was informed of
the accomplishments of Europeans, he retorted that it was true they could
fly and speak across the ocean but they had "no songs like ours, no poets
to equal the island singers" (Grimble, as quoted in Finnegan 1988:83).

When orality and writing coexisted for some time, as in ancient Greece,
some championed one technology while others championed the other,
and still others were ambivalent. Socrates, who lived during the period
when literacy was becoming common in Athens, was ambivalent toward
written languages. He noted that writing had the same relationship to
knowledge as a statue to life. A statue appears to be lifelike but is not
alive. In a similar manner writing appears to contain knowledge, but does
not. He also observed that writing would destroy the ability of people to
remember. Yet Socrates thought it desirable that young men learn to read.
Plato, who followed Socrates by a generation, championed the written
word. In Plato's ideal society poets were to be banned.

Prior to the invention of recording technology, the only procedures available for transforming oral compositions into writing were for the oral specialists to master writing or for an oral specialist to recite a poem to a writer. Many of the oral specialists were not inclined to master writing, and it is difficult for a poet to perform in a manner that allows another to transform oral compositions into writing. "The slow process of writing made the singer lose his nerve, he confused the story, made nonrhythmical verses, dropped into story telling in prose, and tended to shorten the poem" (Jensen 1980:81). Lewis (1978) made similar observations about the South Seas navigators.

When writing first appeared in ancient civilizations, oral technologies continued to be used much as they had been for several generations. In general, orality was used to preserve traditional information and writing was used to preserve information relevant to newly emerging enterprises. As the two sets of specialists managed different types of information in most instances, only small amounts of the orally preserved information were translated into writing as writing replaced orality as the dominant technology.

CONCLUSIONS

Oral specialists in many ancient societies composed and recited poems that served as encyclopedias long before written languages were invented. Those societies with oral specialists were able to preserve more information than those without them. A few of those societies that developed elaborate oral technologies evolved into civilizations.

The nonliterate/literate is a watershed dichotomy but has a large twilight zone. Many complex societies emerged and endured for hundreds, if not thousands, of years before writing was invented, and when literacy did appear it was used only to preserve particular bodies of information. It seems that all preliterate ancient societies with a complex social structure made extensive use of orality to preserve information. The overwhelming bulk of information retained orally has been lost. An analysis of the earliest written accounts indicates that in each instance when literate civilizations emerged, they emerged from civilizations that made extensive use of oral technologies. While literacy did not emerge directly from orality, orality probably was a necessary precursor for literacy.

Orality spiced up the lives of all. Many oral recitations were offered almost entirely as communal entertainment. Usually only a few performers occupied center stage when nonliterate communities held celebrations, but the audiences to the performances were not merely passive recipients. Those in attendance contributed to their own entertainment as they laughed and wept in response to the performance. They thereby affirmed their embeddedness with one another.

Oral recitations often excited people but seldom encouraged the adoption of an analytical standpoint. In contrast, markings of sequences, notches indicating quantity, depictions, and written languages were less exciting but the information could be inspected again and again. The display of information in artifactual form probably was necessary for the adoption of an analytical standpoint and the development of concepts.

Orality is far more than a historical oddity limited to a few ancient societies and some contemporary nonliterate societies. It was necessary for the rise of civilization. Orality along with depictions, calendric systems, and numeric concepts provided the informational base for the establishment of the most ancient civilizations. When orality was replaced by writing, humanity lost some of its vibrancy, but gained the ability to construct more complex social orders.

Orality as a technique for preserving information has been rendered obsolete. However, oral recitations continue to be used to affirm collective sentiments and social unity. The singing of hymns and anthems continues to affirm temple and state solidarities. The advent of recordings and electronic broadcasting has revitalized the significance of evocative communication, but largely for recreational purposes. Very rarely do modern people use an oral technology to preserve information.

3

Decorations and Depictions

Static visual decorations and depictions offer point-in-time configurations that, save for attenuation due to the erosion of material, are constant. In contrast, speech offers across-time ephemeral auditory sequences. When people began constructing decorations and depictions, at least one hundred thousand years ago, they acquired the capacity to communicate with the use of artifacts and constructed the first technologies that allowed them to share experiences in new ways. Prior to the development of that ability all communication was processual and action based.

Decorations display sentiments by offering visual patterns arranged to express and elicit emotions. Purely decorative visuals, such as abstract paintings, do not attempt to replicate other configurations, whereas depictions such as drawings, paintings, and photographs of objects attempt to do so. Depictions represent other objects and carry implications of truth or falsity. They either accurately "re-present" other configurations or they do not. The fidelity of depictions can be assessed, but the fidelity of decorations cannot. Most visual displays both depict and decorate; that is, they are referential and evocative. Decorations and depictions provide a means for preserving and sharing information as surely as do spoken languages.

Decorations and depictions are frequently used as personal adornments to express self and inform others of our character and identities. A host of personal adornments informs most social encounters between strangers in urban centers. We arrange our hair, wear distinctive clothing, and display jewelry to inform others of who we are. Many adornments are so taken for granted that they seldom enter our consciousness in encounters with friends and acquaintances. We are seldom conscious of the wedding ring of a friend but its absence is likely to be noted.

TWO-DIMENSIONAL VISUALS

Drawings, paintings, and photographs infuse nearly all societies. A few groups prohibit the depiction of some configurations. For example, fundamentalist Muslims suppress depictions of the human figure. A few contemporary primitive societies, such as the Pygmies of Zaire, produce only a few simple drawings (Marshack 1989:332). In most societies, nearly all are capable of communicating via drawings, paintings, and photographs. The foundational quality of communicating via depictions is indicated by the fact that most children learn to recognize depictions of configurations and to produce them before they learn to read and write.

Some of the most ancient visuals are configurations in the shape of the human hand. These configurations consist of colored areas that enclose the hand shape. Some aboriginal Australians produced such displays by placing the hand against the wall and blowing a mouthful of colored liquid on the surrounding wall.

The most ancient visuals probably were produced to provide transitory amusement. Human tracings that were the by-products of other activities, such as footprints, may have been the original stimulation for the production of two-dimensional visuals. Ancient people may have been intrigued by such tracings and may have made additional ones to amuse themselves.

The most ancient depictions are crude sketches; the less ancient are more refined and colored. A number of drawings and paintings of animals date from thirty-two thousand years ago (Davidson and Noble 1989:127). Most of the depictions that have survived from that time have been found on the walls of deeply recessed caves. That, however, does not mean that these people sketched and painted only on the recessed walls of caves. Traces of a few open-air etchings from that time have survived. These people probably produced most of their visual displays out of doors on perishable surfaces, such as bark and hides, which have been destroyed by the ravages of time. The drawings on the walls of the recessed caves provide us with some information about what these people thought was significant. We can infer with a fair degree of certainty, for example, that those who drew pictures of bison thirty thousand years ago held strong sentiments toward bison.

Decorations and depictions, upon being constructed, became objects of reflection. The reflective activity associated with the production of depictive visuals may have been accompanied by a transformation in vocal communication. Perhaps depictions were necessary to "transform (vocal) communication from a system of context-dependent signals into a system of reference independent of context" (Davidson and Noble 1989:132). The presence of referential symbols in the speech of contemporary societies that produce very few depictions renders that theory suspect. But conversely, when two people who do not speak the same language attempt to communicate, they often make depictions to facilitate communication. In a similar manner, it is common for adults to use depictions to extend the vocabulary of young children. Depictive configurations may have been a cornerstone of referential thought, but the evidence supporting that proposition is not overwhelming.

The most ancient depictions, such as drawings of bison, re-present configurations derived from mundane experiences. Imagined configurations were not produced until millennia later. The ability to imagine configurations not experienced in the mundane world is contingent on reflecting about depictions that represent mundane configurations. When ancient artists began producing imaginary configurations they sometimes demonstrated a sense of humor. One ancient statuette is of an ibex with a gigantic turd emerging from its rear, upon which two birds are perched. The ability to imagine unicorns, flying machines, and space vehicle rests on a foundation provided by depictions; engineering and science, as well as art, would be impossible without imaginary thought.

Intuition, the ability to arrange events and objects in a particular pattern without being able to specify the arrangement in discourse, also is a derivative of reflecting about visual displays. Intuitive thought pervades such varied activity as young children drawing patterns that they have not been taught and architects designing civic centers.

When human beings established sedentary communities, they began making more elaborate two-dimensional configurations. The walls of the communal buildings of ancient sedentary communities were covered with paintings and reliefs. For example, complex multicolored paintings of partly domesticated cattle that date from about 6000 B.C. cover the walls of Catal Huyuk (Mellart 1967).

When temple- and palace-centered civilizations emerged, the walls of the temples and palaces were covered with paintings and reliefs. These configurations, as is the case for paintings and reliefs associated with modern religious and state centers, were produced to elicit appreciative responsiveness. Specialists in the production of two-dimensional configurations emerged coterminously with temples and palaces. Many citizens of these ancient civilizations decorated their homes with small reproductions of the paintings and reliefs that decorated temples and palaces. Many contemporaries decorate their homes with paintings of Buddha; others with paintings and photographs of heads of state.

Although each society has its own distinctive style, the drawings and paintings produced by primitive groups are not informed by canons of production. The Egyptians of about 2700 B.C. were the first to develop a formula for the production of decorations and depictions. Egyptian paintings and reliefs produced prior to that were more realistic than those produced subsequently. The canons for artistic production of the ancient Egyptians emerged at the same time as or shortly after the Egyptians developed spatial concepts such as the circle, square, and triangle.

The production of decorations and depictions moved from a handicraft mode toward mass production as techniques for producing multiple copies were perfected. The Sumerians of the third millennium before the Christian era used configurations carved in ivory to produce multiple copies of emblems that apparently indicated ownership. The ancient Chinese made extensive use of woodcuts. The woodcut, which was a step toward the production of print, became popular in Europe in the fourteenth century. The earliest popular woodcuts in Europe were depictions of religious scenes such as the suffering of Jesus.

After printing developed in Europe, the woodcut was hybridized into printed broadsheets. The earliest broadsheets were produced by religious zealots and directed at the semiliterate and illiterate citizens during the Reformation. Ironically, these broadsheets contributed to the transformation of Western European culture from an image culture to a print culture, which occurred as the mass production of texts by printing presses eroded the significance of communication via drawings, paintings, and sculpture.

PHOTOGRAPHY

The invention of photography in the nineteenth century revolutionized the production of two-dimensional configurations. Photography allows people to produce configurations that have far greater fidelity to what is represented than drawings and paintings. In comparison to drawings and paintings, the fidelity of photography is truly astonishing and allows even clumsy people to produce replicas of configurations of high fidelity.

The high fidelity of photography enticed some to presume that photographs were untainted by the interests of the photographer. But it soon became apparent that the interests of photographers influenced what is and what is not photographed and that photographs could misinform as well as inform. The maturation of photography was accompanied by greater self-consciousness of how sentiments and interests influence the production of two-dimensional representations of configurations, and artists became more concerned with elucidating particular dimensions of the objects painted and less concerned with fidelity. Among artists, a concern with the emotional impact of paintings replaced a concern with fidelity, as photography became the dominant means for producing images.

Prior to the invention of photography, only the elite had the opportunity to inspect enduring images of themselves. Self-consciousness of appearance became widespread as the camera became a household item. The cliché, "You don't know how you look until you have had your picture took," became popular by the end of the nineteenth century when inexpensive cameras became available. Identities came to be based more on appearance and less on action as photography made identities more visual.

Photojournalism first became a significant procedure for disseminating news during the American Civil War (1861–1865) when newspapers published photographs of bloated bodies and other carnage. "War became a painful, almost tactile, experience" (Feldman 1987:425). Photojournalism rendered war less heroic than romanticized paintings and epic narratives. After World War II, the dissemination of photographs of the German killing centers and their victims exposed the lie of the Third Reich. During the U.S./Viet Nam conflict a photograph of a screaming child who had been sprayed with napalm was widely disseminated and contributed to the growth of antiwar sentiment in the United States. Many similar photographs have exposed some of the horrors of wars. Photographs seem to be more effective in eliciting compassion for the victims of violence than drawings, paintings, oral narratives, or written descriptions.

By the end of the nineteenth century an interest in exotic people and places became widespread, and photographers traveled far and wide to placate that interest. Books and magazines containing photographs of foreigners, mountains, and beautiful people became common in many nations. Conceptions of foreign people and places became more visual. Photographic magazines and books flourished until the maturation of television.

Most photographs are produced for a limited audience, often only for the immediate family or friends. Tourists take photographs to document their travels; big-game hunters to document their kills; wedding portfolios are created to memorialize sentiment and family. Such photographs partially transform ephemeral experiences into enduring tangible documents. Some photographs are taken and distributed to mass audiences to document the occurrence of events. Bolivian authorities distributed photographs of Ernesto Che Guevara's body in 1967 to convince the world that the authorities had prevailed over the rebel.

Those who undertake reflective analyses of visual phenomena routinely use photographs to enhance their observations. Military units use photographs to assess the damage inflicted. Scientific researchers use photographs to accumulate and preserve data. Biologists photograph plants and animals to enhance their visual acuity. In recent decades audiovisual recordings on film and magnetic recordings have partially replaced photography as a procedure for enhancing the visual acuity of researchers, but photography remains a significant procedure for the preservation of data in many arenas of research. The freezing power of still photography remains a powerful technique.

The authorities of state structures and other complex organizations often use photographs to extend their control of subordinates. In 1871 the state authorities of France became the first to use photography as an instrument of surveillance when they used photographs to round up members of revolutionary group (Sontag 1977:5). Many other state authorities quickly followed their lead. Currently most states issue passports, driver's licenses, and security cards that include a photograph. Many organizations issue identity cards to their employees that include a photograph.

Despite the popularity of photography, nearly everyone continues to produce drawings and paintings to express themselves, amuse others, and preserve and disseminate information. Many doodle while attending meetings. Children draw pictures to please their parents. Would-be artists attempt to attract a following by creating distinctive depictions and decorations. Engineers, executives, scientists, teachers, and carpenters routinely sketch configurations to enhance thought and communication.

Ancient manual depictions provided the foundation for pictographic writing; and pictographic writing, in turn, provided the foundation for phonetic writing. Still photography provided the foundation for recording dynamic visuals on film and magnetic tapes.

SCULPTURE

Sculpture offers three-dimensional configurations with a tactile dimension. The antiquity of sculpted figures may be as great as two-dimensional decorations and depictions, but the antiquity of sculpture suggested by ancient statutes may be misleading. Carvings of stone, ivory, and bone are more likely to survive than drawings and paintings on perishable materials.

Sculpted figurines that date from thirty thousand years ago depict females with enlarged breasts and buttocks and lack facial features (Davidson and Noble 1989:136). The absence of facial features may simply indicate that the ability to portray three-dimensional human figures in detail had not been mastered. Some advance the theme that these figurines were icons used in communal ceremonies and reflect awesome sentiments linked to the reproductive capacity of women.

Figurines of pottery antedate pottery dishes and sedentary communities. All sculpted and baked clay figurines that antedate sedentary communities are small. The most ancient large sculptures were adjuncts of the monuments and temples of large sedentary communities. Gigantic sculptures of the state elite, such as kings, do not appear until the beginning of the Middle Kingdom of Egypt, about 2000 B.C., that is, about a thousand years after palace-centered state structures and monarchies had been established. Subsequently, the monarchs of most civilizations called on their citizens to honor them with gigantic statues.

Sculpture often is arrayed around and in temples and palaces to tell a story. The sculpture associated with Christian churches often depicts the crucifixion and resurrection of Jesus. In a similar manner, the sculpture of ancient Sumerians, Egyptians, Chinese, and Maya depicted personalities and events of the past and expressed sentiments toward. Sculpted renditions of the past antedate written histories by millennia. Sculpture was used to offer accounts of the past millennia before written histories were constructed.

The construction of gigantic sculpture of state elites has been more closely affiliated with monarchies and dictatorships than with democratic state structures, but the carvings on Mount Rushmore demonstrate that gigantic sculptures of state elite are not limited to monarchies and dictatorships. Gigantic sculpture usually attempts to legitimize forms of social structures as well as memorialize personalities. The faces on Mount Rushmore attempt to legitimate the American presidency as well as memorialize the personalities of four American presidents. However, sculpture only legitimates and memorializes when contextualized by narrative in history books, brochures, and oral accounts. A sculpture that is not contextualized by discursive information is only a curiosity.

The relevance of sculptures of state elites has often been short-lived. On several occasions people have destroyed statues of their past rulers. The destruction of statues of Lenin and Stalin that accompanied the collapse of the Soviet Union is but the most recent example of people erasing reminders of their past. In contrast to the often short-lived relevance of sculptures of the state elite, sculptures of temple elite often have endured for millennia. Statues of Buddha and Jesus continue to occupy a central position in many religious centers and still decorate many homes.

Sculpture does not occupy the dominant position it once did in the preservation and dissemination of information. Statues honoring the elites continue to be erected, but other forms of communication have partially replaced them as procedures for disseminating information. For example, contemporary heads of state are more likely to appear on television to establish the centrality of their personalities than to have a statue constructed in their honor.

PERSONAL ADORNMENTS

The donning of decorations and depictions is a more complex activity than producing them. The use of personal adornments to communicate one's character and identity is based on reflexive action wherein one adopts the standpoint of another toward oneself. It is unlikely that personal adornments are as ancient as drawings, paintings, and sculpture, but beads and pendants of shells and animal teeth have been found "in all regions of the Eurasian Upper Paleolithic" (Marshack 1989:332).

All adornments that date from Paleolithic times have been found in conjunction with homo sapiens. Of the many burials of Neanderthal man excavated "not a single one was accompanied by person adornments" (Clark 1986:7). Does the absence of personal adornments from the Neanderthal burials indicate that these people were not self-conscious?

Personal adornments are not as ancient as other forms of decorations and depictions, but they are as ubiquitous in contemporary life. It is rare indeed for an adult to appear in public without benefit of personal adornments. In many contemporary societies, adults are more likely to appear in public without benefit of clothing than without displaying adornments. Entertainers sometimes appear unclothed, but rarely do they appear without adornments. In a parallel manner, some ancient cave drawings depict the human form decorated with paint and jewelry, but devoid of clothing. The Ona, natives of the southernmost region of South America, did not cloth themselves, but they painted their bodies (Flugel 1938:16).

Modesty, protection from the elements, and personal adornment are not mutually exclusive. The young woman who appears on Fifth Avenue on a winter day wearing a sable parka shields her body from others and the elements and also offers an adornment. Both the well-dressed woman on Fifth Avenue and the Stone Age hunters who adorned themselves with necklaces of bear claws intended to elicit particular responses from others. When a judge dons a robe and seats herself on an elevated platform, she claims she is a worthy person and that others owe her deference; when an actress appears displaying elaborate makeup, she hopes to elicit appreciative responses.

Many adornments, such as pink clothing for girls and blue clothing for boys, are foisted on children to communicate identities to those who wear them as well as to others. The personal adornments that indicate gender implicitly call for others to behave in a particular manner toward the person displaying the adornments. Young children, of course, are not conscious of either their identities or the responses called for by the adornments they display. Adornments per se do not generate self-consciousness, but they may be necessary for the creation of self-consciousness. To become self-conscious a child must adopt the standpoint by another toward self and designate self with a referential symbol. The emergence of self-consciousness is facilitated by the display of adornments. Reflexive activity, the adoption of the standpoint of another toward self, underlies the donning of all adornments.

The donning of a nurse's uniform by someone who has worked as a nurse for several years typically is not a reflexive act, but rests on a foundation of reflexive activity. When a person is invested with the authority to don a nurse's uniform she typically is very self-reflexive when she dons an adornment that designates her newly acquired identity. Self-consciousness of an adornment designating an identity fades as the identity and the adornment that depicts the identity become taken for granted.

Some adornments, such as decorative ones, attempt to communicate character. When an administrator selects a tie before going to a meeting, the act is structured by his intention to inform others of his character. Depictive adornments call for acknowledgment of one's identity. Many adornments, for example, engagement rings, both decorate and depict.

The character and identities people attempt to communicate via adornments usually are validated by others, but on occasion they are not. The necktie selected to elicit respect, may be responded to with mirth. The importance of personal adornments for social order is suggested by the fact that performing artists, temple and state elite, and political candidates often are served by personal attendants to assure that the adornments they display will elicit an appreciative response from audiences. When the anticipated response is not elicited by an adornment, the social encounter either disintegrates or other dimensions of character and identities are negotiated. The display of adornments that fail to elicit anticipated responses often is a source of embarrassment.

The donning of decorative adornments became more elaborate after mirrors were invented. Cosmetic pots, hairpins, and combs, have been found in association with mirrors dating from 4500 B.C. (Goldberg 1985:31). People frequently use mirrors when donning adornments, especially when they prepare to participate in out-of-the-ordinary events such as weddings. Those without mirrors must don adornments on the basis of their imaginations and the assessments offered by others. Mirrors allow people to adorn themselves in a refined manner without interacting with a copresent another.

Adornments, like spoken languages, are culturally specific. A ring on the third finger of the left hand informs self and others of one's marital identity only in some societies. Not all adornments are embodied; some are freestanding but linked to a person. For example, executive washrooms, beautiful automobiles, and private hospital rooms communicate one's character and identity much as do clothing and jewelry.

Many adornments communicate membership and position in social structures. A policeman's badge declares both membership in a particular social structure and position within that community. During the European feudal period, the wearing of purple declared one's nobility and

the amount of purple cloth that could be worn was specified for each rank
of aristocracy. In most nations it is a crime to display false military
insignia.

In locations where strangers routinely convene, adornments are commonly used to communicate categorical identities. When categorical identities are established via adornments it allows strangers to become
socially situated the moment they visually note one another. When social
encounters are populated by people who have a shared past, personal
adornments usually are of less significance than when strangers encounter one another. One does not normally use adornments to communicate
one's friendship; however, failure to don particular adornments may render some relationships based on a shared past problematic. The married
woman who removes her wedding ring at a nightclub may render her
marital relationship problematic.

A few groups, such as the Amish, discourage adornments that differentiate self from other members of the community. Nonetheless the
adornments displayed by members of such communities communicate
identity as surely as those that differentiate self from all others. When all
display similar adornments, collective identities supersede personal ones.
Some adornments communicate the character and identities of collectivities. The images of lightening bolts worn by the Nazi SS soldiers
communicated the character of their units as well as a collective identity.

Adornments often are donned for particular occasions to communicate
respect for an occasion. The Sioux Indian preparing for the Sun Dance
wore special apparel much as does a college student preparing for a
graduation ceremony. The donning of elaborate adornments, such as
masks, is associated with either playful or extraordinarily serious occasions. Many adornments are used to dramatize occasions. Dressing for a
special occasion often is a source of enjoyment and pride.

Adornments may be a source of either pride or mortification. Those
recently invested with a sought-after identity, for example, newly commissioned military officers, sometimes parade about in public to display
the adornment that designates their new identity. Conversely, branded
criminals usually attempted to hide their stigmatizing markings. Many
regard personal adornments, especially decorative ones, as superficial
artifacts of little significance, but people fight and die for the prerogative
of donning adornments that designate prestigious identities. When societies with different levels of refinement in decorative adornments make
contact, many of those from the society with the less refined personal
adornments seek the more refined ones.

Nearly all social encounters are informed by personal adornments.
Exceptions include social encounters that take place in total darkness and
via some types of telecommunications. Personal adornments, particularly

depictive ones, facilitate becoming socially situated and the construction of complex social structures. The emergence of civilizations was contingent on the development of complex sets of adornments that depict identities.

Adornments also are a source of personal stability. Identities designated by adornments usually are more enduring than those designated only via speech. The mere display of adornments designating categorical identity usually is sufficient, but often categorical identities presented via speech sometimes have to be reenacted before others accept them as valid. Of course, not all identities offered via personal adornments are valid. People can lie with adornments as well as with speech.

HOW STATIC VISUALS STRUCTURE EXPERIENCES

With the exception of photographs, static visuals are produced sequentially. But once produced, they are usually experienced as point-in-time configurations. Static visuals are, in a sense, timeless; they freeze processes and they endure. In contrast, spoken discourse is timeful and ephemeral. When people speak to one another, anticipations structure their experiences; fluidity and change characterize discursive communication. Static visuals implicitly deny flux.

Static visuals sequence neither the past nor the future, nor do they measure duration. We appreciate and assess discursive information largely in terms of the future implied; we appreciate and assess visual displays largely in terms of the present and past. Visuals preserve the past and implicitly claim the past is significant. Museums are the prototypic visual constructions that assert the significance of the past.

Paintings and photographs imply living organisms exist as frozen configurations, but life is dynamic. The presentation of a life form as if it were static misrepresents it. Nonetheless, static visuals that replicate configurations enhance the adoption of a reflective standpoint toward objects and thereby have contributed to extending our understandings of life.

Across-time and point-in-time communication, despite their profound differences, are not antagonistic. Visual displays are frequently used to enhance discursive communication. Teachers often sketch diagrams on blackboards and lectures are sometimes accompanied by a slide show. If communication is restricted to discourse one has great difficulty in communicating the configuration of a platypus to one unaware of the animal. A multitude of words is required to accomplish that objective when communication is restricted to discourse, and even then it would be questionable if the naive person would recognize a platypus when he encountered one. But a photograph of a platypus accompanied with a few instructive words will provide all the information necessary to assure recognition of a platypus when one is subsequently encountered.

Depictive visuals, whether a chalk drawing on a blackboard, an insignia, or a photograph, answer the question: What is it? They are constructed to be recognized. Recognitions allow us to organize ourselves toward our environment. Depictive visuals do not assure mutual understanding, but they greatly enhance the likelihood of communication, especially when novel or complex issues are addressed.

CONCLUSIONS

Our thoughts and sentiments are given form by constructed visual configurations as surely as they are by vocal discourse. The power of configurations to retain information and communicate is suggested by the term *iconoclast*. Iconoclasts destroy venerated images and beliefs.

Decorative visuals contribute to the constancy of emotions. The emotions elicited by static decorations tend to be quiescent; those elicited by speech, activities, and dynamic visuals tend to be volatile. An audience viewing a statue of a religious figure is typically more pensive than one listening to a speech about the religious figure or watching a motion picture that offers a characterization of a religious figure.

Two-dimensional visuals may be a more primitive form of referential communication than referential speech. One of the first information technologies mastered by young children is picture books. Picture books are based on correspondence; with few exceptions, vocal symbols are not based on correspondence. Discursive symbols are arbitrary. The configuration of a drawing of a cat must correspond with the configuration of a cat. The sound of *cat* does not correspond with the configuration of cats nor any feature of cats. The word *meow* is an attempt at correspondence.

Most communication via configurations is intertwined with discursive information. For example, the sand paintings of the Navajo are produced as part of a ritual that includes chanting as the configurations are constructed. Visuals also are closely intertwined with discursive information in contemporary societies. A statue of the Virgin Mary has little significance unless contextualized by discursive information.

Configurations ranging from the Great Pyramid of Egypt to last Sunday's comic strips enrich the lives of all. Our lives would be impoverished without them and life would be far less delightful.

Motion pictures, television, and magnetic recordings are partially supplanting static visuals. Yet despite the prominence of the dynamic visuals offered by recordings and television the visual configurations offered by paintings, still photography, sculptures, and personal adornments continue to be major sources of information and pleasure.

CHAPTER

4

Extending Temporal Structures

Several times each day, except perhaps when on vacation, most of us glance at calendars and clocks. We use the information obtained to organize our conduct. If we note it is only two days until our spouse's birthday, we may go shopping for a gift; if we note it is only fifteen minutes before the convening of the departmental meeting, we may scribble a note or two in preparation for the meeting. Despite the fact that we routinely use the information obtained from timekeeping devices to structure our actions, we seldom give the artifacts that display temporal information or the information itself a second thought. The information provided by calendars and clocks is part of our taken-for-granted world.

If all calendars and clocks and our memories of them were to disappear, all complex social structures would collapse. The continuation of complex social structures is contingent on people programming distal futures; and the programming of distal futures is, in turn, contingent on timekeeping technologies. Only societies with complex timekeeping procedures can create and maintain complex social structures. Timekeeping procedures are foundational information technologies.

In Marshack's (1972) words, time-factoring procedures were the *Roots of Civilization*. Procedures for specifying the seasons of the year—calendars—and using those procedures to program future activity have a very ancient past. Techniques for specifying the time of day are a more recent development, but they too have an extended past. More recently, the development of mechanical procedures that standardized small units of duration and told the time of day provided one of the cornerstones for the emergence of industrial societies.

Timekeeping technologies are not panhuman. Some contemporary nomadic societies endured without benefit of a timekeeping technology. For example, among the Siriono, "The year, with its division into months or 'moons,' is quite unknown" (Holmberg 1969:122). The only temporal terms they used to structure distal futures and pasts were today, tomorrow, yesterday, brother-of-tomorrow, and brother-of-yesterday (pp. 122–23). The Siriono used the term *brother-of-yesterday* to locate events as occurring in the past, but that term only located events as occurring before yesterday. Societies such as the Siriono did not project long-range futures.

Nomadic societies without calendars programmed futures of only limited duration and the future was only imprecisely programmed. In contrast, members of modern societies project precisely program activities that extend far into the future. Modern timekeeping procedures allow us to specify temporal structures in milliseconds. The outer space probes of NASA are but one example of social endeavors that are complexly and precisely programmed.

Temporal structures are integral dimensions of all social activity ranging from the simple act of two people walking down the street together to a nation organizing itself to repel an invasion. The simpler forms of coordinated action are achieved on the basis of interpersonal timing: by people timing their activities with one another. The completion of the complex coordinate endeavors undertaken by large social structures, such as nations, however, requires timekeeping technologies.

BASIC TEMPORAL STRUCTURES

The behavior of newborn infants is not timeful. Infants can neither anticipate events nor act with intentionality. They only respond to stimuli; they do not act. Before infants can organize their behavior, they must acquire the ability to anticipate.

Infants acquire the ability to anticipate by participating in a series of transactions with their caregivers, as discussed in Chapter 1. After a child has participated in a series of transactions wherein his caregiver has appeared in his visual range and then made tactile contact with him, he then can anticipate and prepare himself for tactile contact when his caregiver appears in his visual field. His behavior thereby becomes temporally informed.

Whereas anticipations first emerge on the base of initiations of others toward the infant, intentionality emerges from activity initiated by the infant. The simplest intentions stem from the random acts of the child, such as the child flailing his arms. For example, the infant may accidentally bring his hand in contact with his mouth. That may stimulate a sucking response. After such an event occurs a few times the infant may then intentionally move his hand toward his mouth to suck his hand. The earliest anticipations and intentions are of short duration, but as children mature and become mobile their temporal structures are extended. Human actions ranging in complexity for a person scratching himself to relieve an itch to a nation mounting a campaign to rid itself of a deadly disease are temporally structured.

Intentional action presupposes anticipations. If a person cannot antici- pate the future he cannot formulate an intention. When a person cannot formulate an intention he can only respond to stimuli; he cannot act toward his environment.

In a similar manner, if a collectivity—ranging from a dyad to a na- tion—cannot formulate shared anticipations and project a shared future, coordinated action is an impossibility (Hintz and Couch 1975). People can act with respect to one another without projecting a shared future, but to act with one another—in unison—requires that they project a shared future and then align their actions with one another. The formulation of shared anticipations and intentions does not assure that a collectivity will achieve the objectives sought, but shared anticipations and intentions must be formulated if coordination of action is to be produced.

Since it is impossible to anticipate all future developments, the produc- tion of coordinated action requires that people be responsive to unfolding events as well as anticipate future events and formulate intentions. Ac- tion, both personal and social, is the consequence of the relations between projected futures and unanticipated developments. As young children have only limited anticipations, they can only exercise limited control over their experiences and actions. Human beings acquire greater control over their lives as they acquire command of complex temporal structures.

Much coordinated action produced by enduring groups is informed by shared pasts. Each time a basketball team practices or plays a game, they elaborate or entrench their shared past. When the members of a to-be- basketball team convene for the first time to practice, they do not have a shared past. Typically each member of a to-be-team has a past of having previously played basketball; they have common pasts. Common pasts as well as shared pasts can be used to organize social action, but when people do not have a shared past usually they must explicitly sequence their activities if coordinated action is to ensue. In contrast, when people have a robust shared past, they can use taken-for-granted shared tempo- ral structures to organize their actions. For example, an emergency crew can instantaneously produce complex units of coordinated action the in- stant the shared focus of an emergency is established.

Elementary shared pasts such as those constructed by a presymbolic infant and her mother allow for the projection of primitive shared futures. But the production of coordinated action of any complexity requires mas- tery of symbolically sequenced temporal units (Couch 1984:37–38). It is only after a child has acquired command of symbols such as first, next, after, before, etc. that a child can symbolically sequence her action with others. A primitive level of symbolic sequencing is achieved when the child requests the parent to do something and the parent responds with "just a minute" and the child maintains an anticipatory state. In such transactions the child projects a future, and the parent temporarily denies the projected future, and the child as well as the parent uses the projected future to structure their conduct.

Taken-for-granted shared pasts structure such varied activities as a shopping expedition, a birthday party, and playing a basketball game. Usually we only become conscious of the temporal structures that inform routine actions when attempting to instruct a neophyte into the intricacies of the activities or when the activities are not produced in the sequence necessary for the completion of the activity.

Fairly complex units of social action can be brought to successful fruition without benefit of a calendar. One member of a band of food gatherers may, upon noting the presence of game, alert others. Then all adult members of the band may organize themselves for a hunt. If they sequence their actions with one another and with the movement of the prey, and all agree on how they are to proceed, a fairly complex social endeavor can unfold. In such instances one person may take charge and direct the actions of others. The organization of the action may be no more complex than all mutually agreeing that some of them will circle behind the quarry and drive them toward the other members of the band. In such instances the action involves the projection of a social objective (killing the game), interpersonal timing (each attending and responding to the others during the hunt), and timing their actions with changes in the external environment (adjusting to movement by the quarry). The completion of units of coordinated action of that complexity requires the presence of some who have the ability to offer rather elaborate symbolic sequences. Coordinated action of this sort pervades the lives of all societies. The anticipations and intentions that inform such units of action are linked to proximal futures, objectives to be achieved in the immediate situation.

SIMPLE CALENDARS

Modern calendars and clocks allow us to program futures such as "in two months, on June 19 at three in the afternoon we will convene and hold a seminar on the temporal structures that inform modern states." Such planning is beyond those who do not have command of a numeric calendar and clock.

The evolution of timekeeping procedures that allow for such activity spanned tens of thousands of years. People began to develop calendars when they began noting that certain significant events such as the reappearance of migrating herds of animals followed the occurrence of other specific events. As people began using information derived from such observations to program their future activities, they took the first steps toward inventing calendars.

Long before the invention of calendars people must have noted that some terrestrial events occurred in sequences. For example, it might have been noted that after the snow disappeared from the floor of the valley, flocks of geese arrived. Ancient calendars emerged when people began noting terrestrial changes and linking those changes to celestial events, such as the disappearance and reappearance of a bright star.

A major transformation of temporal structures occurred when people invented lunar calendars. Lunar calendars divide the yearly cycle into lunar cycles. At the end of each lunar cycle, the moon disappears from the sky for about two and a half days. There are approximately twelve and one-third lunar cycles in each yearly cycle; consequently there are twelve new moons some years and thirteen other years. Lunar calendars must be keyed to a yearly event if they are provide a procedure for programming activity for the terrestrial seasons. The necessary realignments of lunar cycles to seasonal changes were achieved in a number of ways.

The most ancient lunar calendars were, in all probability, keyed to a terrestrial event. For example, one society keyed their lunar calendar to the first clap of thunder that indicated the rainy season was about to start. They began a new set of lunar cycles each time the rainy season started. Another group keyed their lunar calendar to the migratory patterns of flying fish. Each year after twelve moons had passed, they migrated to a location on the seashore to begin their yearly fishing expedition. Some years the group arrived at the location before the fish. When that occurred they blamed the fish for being late, but they had the good sense to extend the fish-harvesting season by one moon (Leach 1954:119). They thereby realigned their lunar calendar with terrestrial sequences.

Somewhat more refined lunar calendars were keyed to celestial events. Many lunar calendars of the northern hemisphere were keyed to the winter solstice. Others were keyed to the helical risings of a major star. Each star, except the circumpolar ones, disappears from the sky each year. After a star has disappeared it first reappears in the eastern sky just before sunrise; its reappearance is its helical rising.

Those societies that used simple lunar calendars could not anticipate and formulate intentions with the precision that those with a numeric calendar can, but they could formulate anticipations and intentions of far greater duration than can those without a calendar. Those with only a lunar calendar could not formulate such futures as, "On the seventeenth of March we will load the wheat to be shipped to Japan." But they could formulate futures such as, "It is the time of the wet moon, therefore we should go to the seashore and catch fish."

Some ancient groups made marks on pieces of bone and stone to note the appearance of each new moon at least thirty thousand years ago (Marshack 1972). The earliest calendars were a simple series of scratches. They were similar to the marks made by a prisoner in solitary confinement who wishes to keep track of the days. The increase in the amount of information at the command of those who first made calendars was small, but the information allowed for the emergence of somewhat more complex social structures.

The development of lunar calendars was contingent on people making marks on pieces of wood, antlers, and bones that recorded the passage of each moon. Information retained in ordinary speech or orality is too fluid to allow for the development of temporal structures that extend several months (moons) into the future. Even when a person who has command of a numeric calendar is placed in solitary confinement, it is necessary for him to make inscriptions if he is to maintain an accurate record of the duration of his confinement.

The earliest calendars did not measure duration. Societies with simple lunar calendars did not count lunar cycles. Instead they gave names to each moon so that the year consisted of a sequences of names, much as our year consists of the sequence of January, February, March, etc. The names given to each moon in nomadic societies reflected either climatic conditions or human activity associated with each moon. For example, the moon following the winter solstice might be called the frozen moon, the moon that roughly corresponded to our month of April might be called the wet moon, and the moon that roughly corresponds to our month of May might be called the greening moon.

Calendars emerged out of practical concerns and provided information that facilitated the resolution of practical problems. The earliest calendars were used to predict the return of migrating herds of animals, the spawning of fish, and the termination of pregnancies (Marshack 1972). The societies that used these calendars were more likely to survive than the groups that did not have them. These calendars did not assure survival; nor did the failure to develop a calendar mean that a society would fail to endure.

Some lunar calendars continue to be used. Many of the Jewish faith continue to program their temple-centered celebrations with a lunar calendar and the Chinese continue to celebrate the new year on the basis of a lunar calendar.

REFINEMENTS

Most contemporary nomadic societies have at least a lunar calendar, which, in addition to Marshack's research on ancient calendric systems, suggests that lunar calendars were developed long before sedentary communities were established. It is likely that the establishment of sedentary communities was contingent on the prior development of timekeeping technologies that allowed people to project a yearly cycle.

After sedentary communities were established, calendars based on alignments instead of lunar sequences were developed. Before a society could use the locations of the sun and stars on the horizon to time-factor the year it was necessary to be sedentary. The sun and stars appear to rise and set at the same location on the horizon only if observations are made from a constant viewing point. The specification of the rising and setting points of the sun and stars with precision could not have occurred until after sedentary communities were established.

Whereas the older and simpler lunar calendars were keyed to either a yearly terrestrial or a celestial event such as the helical rising of a star, the calendars of the sedentary agricultural communities became keyed to specific alignments created by the risings and settings of the sun and stars. For example, among the Zuni a sun chief watched the shadows cast by a mountain, a solar monolith, and a pillar in the communal gardens. When the shadows cast by these three objects fell in a straight line, the sun chief announced to the community that it was time to begin field work (Cushing, as cited in Ellis 1978).

The development of an agricultural way of life was contingent on the prior or simultaneous development of a calendar. "Groups that do not have a fairly accurate calendar may attempt to become farmers, but their chances of success are minimal" (Couch 1984:156). As agriculture replaced hunting and gathering as the primary source for the necessities of life, calendars came to occupy a more central position in the lives of people and the calendars became more precise. (The names given to each moon in agricultural societies were linked to agricultural concerns. For example, the ancient Sumerians named each moon after the predominant agricultural activity taken during the moon.) The development of agriculture did not cause these communities to develop more elaborate and precise calendars; nor did the development of more precise calendars cause the development of agriculture. Instead the elaboration and refinement of each facilitated the elaboration and refinement of the other.

Most ancient agricultural societies constructed observatories to refine their calendars. The earliest observatories were sticks and stones arranged to mark locations on the horizon where the sun rose and set at different times of the year. The observatories of these ancient communities were centers of attention; all recognized the necessity of using the information provided by these calendars to plant their crops. Communal celebrations were held in conjunction with specific celestial events. The beginning of the yearly cycle—the arrival of a new year—was a major festival day for all societies with a lunar calendar keyed to the celestial event. For example, societies that keyed their lunar calendars to the winter solstice celebrated the winter solstice.

Some ancient agricultural societies transformed their observatories into monuments and many of those attached temples that housed their timekeepers to their monuments. Many of the prosperous agricultural communities built gigantic monuments and elaborate temple complexes. The Great Pyramid, Stonehenge, the ziggurats of Mesopotamia, the temples of the sun and moon of ancient Mexico, Monks Mound near St. Louis, and other monuments aligned to celestial events were preceded by simple observatories. As these communities prospered, they designed monuments to offer dramatic displays of celestial events. One of the better known monuments, but certainly neither the most ancient nor the largest, is Stonehenge, which offers a display of shadows cast by the rising sun at the winter and summer solstice. One of the major monuments of the Maya was designed so that at sunrise on the days of the equinoxes a "snake" appears to emerge from the top of the monument and crawl down the corners of the pyramid. The snake is a series of undulating shadows. The shadows on the corners of the pyramid were and remain an impressive sight.

Some of these ancient monuments were architectural masterpieces. For example, the Great Pyramid of Egypt was originally covered with a layer of polished marble that reflected the sun, moon, and stars. It also had a capstone that reflected the rays of the sun just before the sun appeared on the horizon. Monuments aligned to create particular reflections and shadows at specific times of the year evoked emotions as well as provided temporal structures. They must have been regarded as marvels when first constructed, but after they had been in place for a few years they perhaps became taken for granted.

These monuments served as both timekeeping devices and community centers. They allowed all members of these societies to witness key celestial events and participate in celebrations that reaffirmed communal solidarity. In some ancient agricultural societies thousands of people gathered at the monuments to celebrate the events. Christmas stems from an ancient celebration held at the time of the winter solstice. At each solstice the sun appears to rise and set at the same location on the horizon for a few days. The original meaning of the solstice was "standstill." As the sun does not appear to the naked eye to move northward on the horizon until a few days after the solstice, Christmas originally marked the day when it became apparent that the sun had began moving northward on the horizon. Easter stems from a fertility celebration that marked the first full moon after the spring equinox. Easter rabbits and Easter eggs originally were fertility symbols.

The development of calendars precisely keyed to such events as the winter and summer solstices and the equinoxes preceded the establishment of centralized governments. When the ancient civilizations of Mesopotamia, Egypt, China, MesoAmerica, and South America appear in the historical record, complex calendars and precisely aligned monuments were already in place. Calendars were cornerstones for civilizations, rather than the consequence of civilizations.

These ancient calendars did not quantify time. Numeric calendars, calendars that numbered the days in the year, did not appear until after centralized states were in place. The first numeric calendar was invented about 2800 B.C. two or three centuries after the Egyptian nation was formed. Several ancient civilizations developed numeric calendars. Many of them, such as the calendars of the Maya and the Inca, fell into disuse when the civilizations were destroyed by invaders. Others, such as the ancient Egyptian numeric calendar from which the modern Western calendar is derived, continue in use. Some discussions of these ancient timekeeping monuments treat them as interesting epiphenomena of little consequence, but these timekeeping instruments and the information they preserved and displayed were the foundation upon which the social structure of ancient civilizations rested.

TIMEKEEPING SPECIALISTS

The first information specialists who were provided with necessities of life by their communities were the timekeepers of agricultural communities. There was probably some specialization that centered on timekeeping among some nomadic groups, but the degree of specialization was not great. In contrast, in many ancient agricultural civilizations an elite category of information specialists appeared. These specialists in charge of the calendar and celebrations linked to communal monuments were focal persons in the social structure of their societies.

Exactly how these first information specialists emerged and their exact position in the social structure of their societies cannot be determined with certainty. But it is fairly certain that their elite position was based on their command of information that was critical to the welfare of their communities. Only the information specialists had the knowledge necessary to design the monuments effectively and conduct the celebrations linked to them.

As the monuments and celebrations became more elaborate, greater differentiation between the timekeepers and other members of their communities emerged. One example of that development is the growth of monument construction in ancient Britain. The first agricultural communities were established in ancient Britain about 4000 B.C. (MacKie 1977a). Shortly after an area was settled, the communities began building monuments and distinctive burial chambers. Those buried in the special burial chambers did not do manual labor. The skeletons found in these chambers do not have "any deep grooves caused by the attachment of particularly strong muscles" (p. 151). The teeth of these skeletons indicate that those buried there had had a diet that was relatively free of gritty material. There are also indications that these specialists wore woven clothing instead of the hides worn by others. It seems that the timekeepers of these communities were supported by communal contributions.

Whereas the earlier observatories were designed to sharpen observation of celestial events for timekeeping purposes, the later monuments were designed to elicit awe. As the monuments increased in size and the celebrations became more elaborate, those responsible for timekeeping and organizing the communal celebrations became transformed from men of knowledge into priests who lived in temples apart from the community. In many cases the procedures for noting and preserving information derived from the observation of celestial sequences became secrets known by only the information specialists.

In some regions, complex temple-centered social structures were established prior to the establishment of state structures and the invention of writing. In at least some instances, such as the Incas and residents of ancient Britain, large quantities of information about celestial phenomena were preserved and transmitted from one generation to the next by merging an oral technology with observatories and monuments. Some but not all of these societies subsequently evolved into societies with a state structure.

After palaces and monarchies emerged, the timekeepers continued to occupy a critical position in the social structure of ancient civilizations. Such was the case for Egypt, China, Mesopotamia, Inca, and the Maya. Imhotep, the designer of the Step Pyramid constructed by the ancient Egyptians, was the head priest of Egypt and had the title "chief of the observers." The great pyramid of Giza was originally named "the place of the sunrise and sunset" (Davidovits and Morris 1988:182) and was surrounded by temples that housed priests.

Monuments that displayed timekeeping information and timekeeping specialists continued to occupy critical positions in the social structures of ancient civilizations for thousands of years. For example, gigantic monuments aligned to celestial events were constructed throughout ancient Egypt's three thousand years of existence. The temple of Abu Simbel, built during the Middle Kingdom, was celestially aligned. The temple contains six chambers. The front of the first chamber is over 150 feet wide and its passageway is flanked by gigantic columns. It is over 250 feet from the front of the temple to the back wall of the sixth and innermost chamber. Each of the inner chambers becomes progressively smaller. On the back wall of the sixth chamber is a relief that displays the gods Isis, Horus, and Seth. It also displays Rameses II, who was the king of Egypt when the temple was constructed. The temple is aligned so that twice a year at sunrise the rays of the sun flash down all six chambers and bathe the back wall in blinding sunlight. A further refinement is that only the gods Isis and Horus and King Rameses are bathed in the sunlight. Seth, the god of darkness, remains in darkness.

The priests in charge of timekeeping programmed and supervised the celebrations held at Abu Simbel. The sixth chamber, like the most recessed chamber of all the major temples of dynastic Egypt, was referred to as the holiest of the holies. Only the head priest and the king could enter that chamber; and the king could only enter after he obtained permission from the head priest.

Gigantic monuments, such as Abu Simbel, are usually interpreted as being constructed at the behest of the palace elite. And, in many instances they were. But the temple elite who were in charge of timekeeping and associated celebrations not only provided the knowledge necessary for the construction, they also organized the rituals and more importantly provided temporal structures for the nation.

The celebrations linked to these monuments enhanced communal solidarity in much the same way as celebrations of the Fourth of July enhance national solidarity in the United States. But, as the rituals became more elaborate, the social distance between the timekeepers and the rank and file increased. In many ancient societies as the information specialists become transformed into priests, the monument-centered rituals get transformed from joyous celebrations into burdensome obligations. In some instances the palace and temple elite formed coalitions. These coalitions were not necessarily formed with the intention of more effectively exploiting the rank and file, but in most instances that was one consequence. The palace elite controlled the coercive forces and the temple elite had a monopoly of knowledge on timekeeping procedures and communal ceremonies.

The establishment of monopolies of knowledge and the differentiation of the rank and file from the timekeeping elite were the consequence of both unintentional and intentional activities. As the calendars and celebrations became more elaborate, more time and effort were required to master the information. Simultaneously the information specialists became the focal points of calls for accounts. Since they instructed the community when to plant crops, if the crops were bad, the timekeepers were blamed. Some of the information specialists concluded that if others were kept uninformed the calls for accounts would be less severe, and so a policy of secrecy emerged among the information specialists in the most ancient agricultural civilizations. The earliest written records indicate that secrecy was an official policy among the ancient timekeeping specialists of the Egyptians, Chinese, Inca, and Maya when those societies entered the historical period.

The timekeeping specialists also had command of information that allowed them to make predictions the others could not make. Some advanced the claim that they had power over heavenly bodies. To demonstrate their control over heavenly bodies, they would command the sun to stop its southward movement at the time of the winter solstices.

The early timekeeping specialists probably were primarily interested in preserving and sharing information; and, the monuments originally were designed to communicate—to make information available to the community. But in many instances in the following generations the monuments became instruments that the information specialists used to hoodwink the general populace. These ancient timekeepers probably were the first set of elites who deliberately manipulated information for personal benefit.

QUANTIFYING DURATIONS

The ancient calendars did not measure durations. Quantitative measures of duration were not formulated until after numeric concepts had been invented and applied to timekeeping concerns. The ancient Egyptians invented the first numeric calendar about 2800 B.C., when they divided the year into twelve months of three weeks each and made each week ten days long. The five remaining days were made festival days. Prior temporal structures of extended duration were composed of nominal sequences, but they did not specify the length of durations: Those who used such calendars could not tell how many moons a year contained. They could only name the moons that composed the year.

Numeric calendars transformed temporal structures from nominal sequences into quantitative units. When numeric measures of duration were invented, they slowly replaced the more ancient timekeeping procedures. In most instances they coexisted for hundreds of years with sequential procedures of timekeeping. Only a few civilizations, in addition to the Egyptians, replaced their lunar calendar with a numeric calendar. The Chinese, for example, developed numeric procedures for ordering the future and past, but continued to use lunar sequences for some purposes.

Julius Caesar imported the Egyptian calendar into Europe. The new calendar, of course, did not transform the day-to-day life of Roman citizens. Most of them continued to use a lunar calendar to structure their lives. They planted their crops, for example, by a lunar calendar, not by the state calendar. After the collapse of the Roman Empire, the intellectual elite of the Catholic church began using the Roman calendar to count the passage of years from Jesus' birth to the present. One of the early schisms between the Eastern Orthodox church and the Church of Rome was a bitter quarrel that centered on the specification of the dates for Jesus' death and resurrection (Boorstin 1985:8). Only a very few other than clergy concerned themselves with the issue. Farmers, merchants, and others continued to use some variation of a lunar calendar to structure their lives until a few hundred years ago.

The Gregorian calendar, which most of us equate with the modern Western calendar, is but one of many. It is a modification of the Roman calendar adopted by the Catholic church as the official calendar at the Council of Trent in 1582. The Gregorian calendar slowly became the timekeeping instrument of European nations. After the invention of the printing press almanacs became popular. These almanacs were the precursor of our modern calendar. As almanacs became household items, the monopoly of knowledge centered on timekeeping was destroyed and timekeeping became less of a communal enterprise. It was no longer necessary to hold a communal celebration in conjunction with the winter solstice to inform all that a new year had arrived.

The establishment of a numeric calendar abstracted timekeeping procedures from communal ceremonies and partially transformed collective experiences into units of duration that are experienced as phenomena that have an existence independent of social relationships. Time thereby became a series of discrete units instead of flowing emotional experiences.

CLOCKS

Just as sequential terms were used to order the yearly cycle long before durations were quantified, sequential terms were used to specify the daily cycle long before numbers were used to specify the time of day. Daylight and dark are used in all societies to sequence daily activities. All societies also used terms that are the equivalent of "at sunrise," "during the middle of the day," and "at sunset" to divide the day. Such terms do not measure duration, they only provide a sequential structure.

Most societies had terms that specified durations. The Kaguru, for example, used one term to refer to durations longer than a day and another for durations less than a day (Beidelman as cited in Maines 1989:112). Other terms, such as quick as a wink and in the frying of rice, have been used to specify durations. These terms measure duration in relative terms, but do not standardize durations.

The day was divided into twenty-four units by the Egyptians about the time they invented a numeric calendar. They selected twenty-four stars at about the celestial latitude of Sirius that were about equidistant from one another to divide first the night and then the day into twelve units. The Maya divided the day into two sets of nine units each: one set of daylight hours and another set of nighttime hours. Many societies used sundials, hourglasses, and water clocks to specify units of duration.

The ancient Greeks introduced the practice of dividing the day into twenty-four hours into Europe after they began trading with the Egyptians. The Greeks also developed mechanical timekeeping devices as they began to formulate principles of astronomical movement. The Greek astronomers constructed celestial globes to display the relative location of the Earth, moon, sun, planets, and stars. These celestial globes, of course, displayed all heavenly bodies as moving around the Earth. Some of these globes included a small dial, usually located toward the bottom or to one side of the celestial globe, that "told the time of day."

Mechanical dials that told the time of day disappeared from Europe following the collapse of the Roman Empire. They were reintroduced to Europe when the Muslims attempted to convert the Christian world. Several elaborate and finely wrought devices that told time were constructed at the behest of the political elite in the tenth and eleventh centuries of the Christian era.

The Chinese constructed a complex emperor's clock in 1090 A.D. to amuse the emperor. That clock was destroyed when a new emperor came into power in 1094 (Boorstin 1985:61). Clocks were reintroduced into China by Jesuit missionaries in the seventeenth century. In the 1760s the imperial palace was filled with timekeeping devices of all sorts. As late as the mid–nineteenth century, clocks were seldom seen in China, "except in public offices where it was not uncommon to see as many as a half-dozen clocks in a row" (p. 64).

In the fourteenth century, some European communities began constructing communal clocks. The early European clocks were often located in the spire of the community's church. They, like the more ancient monuments that displayed the time of year, were communal timepieces. The early European clocks only had an hour hand and often gained or lost fifteen or more minutes each day. Usually someone was appointed to correct the communal clock each day. A sun dial was used to determine the correct time.

These clocks were used to mark the opening of the market, the beginning of worship, and sundry communal activities. In 1390 the king of Paris issued an edict that civic officials, markets, and churches were to regulate their activity by the clock of the palace. In Europe it became common practice to schedule events in terms of the hour of the day as indicated by the community's clock instead of at sunrise, high noon, or at sunset.

Clockmaking and then watchmaking became minor industries in Europe. Manufacturers competed with one another to make more accurate and smaller timepieces. In 1786 the London clockmakers reported that they exported about eighty thousand clocks each year (p. 69). Wealthy homes acquired clocks, and wealthy individuals acquired watches. Simultaneously the numeric calendar became a taken-for-granted item in many homes. Temporal structures moved from a collective context to private worlds. Each community had its own timepiece, and many homes had their own calendars and people their own watches. The level of fascination of Europeans with these instruments is indicated by the term *watch*, which is derived from the practice of early watch owners of continually watching their timepieces.

The perfection of mechanical measures of duration allowed for the emergence of a new form of production: the assembly line. Assembly line production requires that each worker subordinate himself to the temporal sequences offered by a machine. When an assembly line functions smoothly, there is no interpersonal timing nor do the workers project shared futures. Instead each person times his actions with the movements of a machine and projects a line of action that is identical to the one just completed. While on the assembly line, each worker lives a solitary and highly repetitive life. The precise repetitive sequences of actions that assembly line workers must produce are efficient, but extremely boring. Most assembly line workers become adept at daydreaming to maintain their sanity.

In the United States until the 1880s, each city specified the time of day for its residents. In most communities twelve o'clock noon was equated with the moment when the sun was at its zenith. Each day the residents of the city of New York were informed that it was high noon by a large ball sliding down a tower on top of a tall building. That procedure for informing citizens of New York that noon had arrived operated through the first decade of the twentieth century.

As intercommunity trade increased and railroads became the major procedure for transporting goods, an interest in standardized time zones emerged. The globe was divided into 24 units and an international date line was established. Growth in intercommunity trade stimulated and flowed from the creation of standardized time zones.

As numeric calendars and clocks have become more pervasive, timekeeping has become more privatized. Prior to the development of standardized timepieces, each person used his community's timekeeping instrument to temporally structure some of his actions. When all have access to a personal calendar and watch, it is no longer necessary to associate with others to temporally structure one's action.

CONCLUSIONS

All social action is structured by projected futures. If a future cannot be projected, coordinated action is impossible. We more or less continuously assess our actions on the basis of the futures we project. We assess objects, events, and persons in terms of an envisaged future. Are they instruments or obstacles to our desired futures?

The significance of calendars for the organization of experiences and actions in modern societies is indicated by a common test for insanity wherein a person is asked if he knows the day of the week, the day of the month, and the year. The sanity of those who cannot provide the correct answer is suspect, because they are presumed to be out of touch with reality. Those who refuse to or cannot employ the established measures of time to structure their lives are judged incompetent, deviant, or both.

Calendars and clocks inform even the actions of those who do not understand them. Young children are sent to school five days of the week at 8:30 A.M. by their caregivers. Those too disoriented to care for themselves are fed and bathed according to the information displayed by the clock.

The calendar is one of the most important information technologies ever devised. Managers of companies, administrators of bureaucracies, and teachers use it to organize action. A professor looks at the calendar to determine how many more class meetings there are before the semester ends. On the basis of the information obtained she may decide to omit the lecture on timekeeping procedures.

The modern hunter by noting the date displayed by the calendar anticipates that the Canadian geese will be arriving in a week and makes arrangements with a friend to build a duck blind the coming weekend. Ancient hunters with a lunar calendar anticipated the arrival of Canadian geese when the moon of "browning leaves" appeared and migrated to the lake to snare a few geese. The hunter without command of any calendar has no awareness of when the Canadian geese will arrive and cannot organize himself to act toward them until he sees them arriving.

The information displayed by calendars and clocks is seldom reflected upon, discussed, or negotiated. It is simply used. In contrast, in ancient societies that used nonnumeric calendars, people commonly discussed, debated, and sometimes negotiated the time of year. Few modern citizens consider the possibility of negotiating the month, day, or year. Nearly all ambiguity and uncertainty about the time of year or day has been removed. Each of us subordinates himself or herself to the calendar and clock. Some of us when on vacation partially free ourselves from the constraints of our timekeeping instruments, but even then we use them to inform ourselves when our vacation is over.

If, by some magic, all citizens of modern societies lost the ability to use the calendar and clock, they would be incapable of maintaining universities, governmental offices, industrial organizations, or international trade. They would still be able to coordinate their actions with each other in the here and now; but their ability to formulate complex programs of action would disappear.

Calendars began and continue as memory aids. The first calendars slightly enhanced human memory and the ability to anticipate and program the future. Those with command of a numeric calendar have far more elaborate and precise pasts and futures than those without one. Calendars in conjunction with writing transformed the past from vague nonsequential accounts of prior events into precisely ordered chronologies. Calendars were a necessary precursor to the rise of historical consciousness.

The original temporal units were day, moon, and year. Millennia later these temporal units were supplemented by or replaced with the temporal units of week, month, solar year, sidereal year, decade, century, and millennium. All adults use the information preserved by the calendar when they date a letter, plan a trip, prepare for planting crops, or schedule a conference. Each daily newspaper displays the date. If we wish to but cannot determine the year or hour of the day, we become momentarily confused. Personal and social order in modern societies rests on information provided by our timekeeping instruments.

CHAPTER

5

Numeric and Spatial Concepts

Numeric and spatial concepts are taught to school children as if those concepts had an existence independent of human action, and many regard them as having truth value independent of their use by human beings. Nonetheless, numeric and spatial concepts are abstract symbols developed by human beings in transactions with their environment as surely as are other symbols. In fact, a number of primitive societies did not have either numeric or spatial concepts.

The only symbols available for characterizing quantity in some primitive societies were words equivalent to "a pair," "a few," and "lots." Some primitive groups had several symbols that referred to pluralities that were content linked. The English language still retains some traces of this practice. We commonly speak of a married couple, a pair of socks, a span of mules, and a brace of pheasants. The symbols couple, pair, span, and brace share the abstract quality of "twoness" for those who have command of numeric concepts, but twoness is not a dimension of the consciousness of those who used content-specific words to reference quantity. The Gilyak had twenty-four terms for pluralities of two (Diakonoff 1983:88), but they did not have the concept two.

Many primitive societies managed quantitative information by matching pebbles or sticks with other items. Members of a society in Sri Lanka used sticks to stand for other objects such as coconuts. When asked how many coconuts they had, they could only point to their pile of sticks and say, "that many" (Menninger 1969:33). In some primitive societies it was common practice to note quantities by making a mark for each item. Matching procedures probably have a more ancient past than one-for-one markings on surfaces, but matching procedures did not provide the foundation for the development of numeric concepts. It is probable that numeric concepts were originally developed by people who made one-for-one markings on wood, bone, clay, or walls.

The arrangement of some ancient markings suggests early quantifiers used their fingers to note amounts. A bone dated twenty thousand years old has fifty-five notches in groups of five. The literal translation of the word *five* for many primitive societies is "a hand." Despite the fact some of the earliest quantitative inscriptions reflect the five digits of the hand, it is doubtful that counting with fingers directly contributed to the formation of numeric concepts insofar as the display of fingers—like the spoken word—is fluid. The development of numeric concepts, however, was contingent on preserving quantitative information in artifactual form.

Both numeric and spatial concepts emerged from practical concerns wherein people preserved information in artifactual form. The first steps toward developing numeric concepts were taken when people began to make markings to preserve quantitative information about material resources; the first steps toward developing spatial concepts were taken when people began to preserve information about the location of celestial objects on the horizon. These markings, like ancient depictive visuals discussed in the previous chapter, preserve information in artifactual form, but whereas constructed configurations referenced concrete objects, numeric and geometric terms are abstract symbols that specify quantity and space.

EARLY NUMERIC CONCEPTS

One-for-one correspondence techniques are symbolic, but they are not conceptual. Numeric concepts did not emerge until after people had used one-for-one techniques to preserve quantitative information for thousands of years. The first numeric concept probably was "two."

The first numeric concepts seem to have emerged in western Asia in conjunction with the management of information about commodities (Schmandt-Besserat 1988:141–44), and were refined and elaborated as sedentary societies traded, accumulated, and stored goods. These communities developed a system of tokens to reference different kinds of commodities. The tokens were made of clay and took several different shapes, with each shape referencing a specific commodity. One type of token represented "one bushel of grain," a token of a different shape represented "one jar of oil," a token of still another shape represented "one pot of beer," and so forth (Schmandt-Besserat 1981:322). This system was not abstract but rather was content bonded, and different tokens were used to count different items. The token system matured in agricultural communities and was developed by specialists in charge of communal goods. The tokens became common in conjunction with "a quantum leap in quantity of cereal pollen in the soil" (Schmandt-Besserat 1989:30), which suggests that the development of numeric concepts occurred in conjunction with an expansion in cereal production. The practice of centralized storage of hard grains may have been a cornerstone for the development of the first numeric concepts.

One finding at Jarmo, dating about 6500 B.C., yielded "1153 spheres, 296 disks, and 106 cones" (Schmandt-Besserat 1982:4), which indicates the rather widespread use of these tokens. These "tokens evidence little change in shape, size and manufacture" for about four thousand years after they first appear (p. 4). Most of the early depictions referred to food commodities, usually grains and animals (Schmandt-Besserat 1981:340), but the amounts of goods referred to were quite small. "The number of animals usually range between one and ten" (ibid.).

"In the fourth millennium, envelopes were invented to hold tokens" (p. 323). These envelopes were hollow clay balls about the size of the human fist. As these envelopes came into common use, people began making markings on the sides of the envelopes to represent the number of tokens inside the clay balls. "It was soon realized that the envelopes filled with tokens and bearing signs could be simplified to lumps of clay bearing signs" (ibid.). People began to make one-for-one correspondence markings on the lumps of clay to specify quantity and drawing pictures on the lumps of clay to designate type of goods. One set of notations specified quantity, while another set specified quality.

The separation of symbols that designated quantity from those that designated quality allowed for the formation of numeric concepts, since the numeric inscriptions were no longer content bonded. As Schmandt-Besserat explains, "The combination of wedges and circles with adjoining pictographs indicating all possible commodities, suggests that these signs took the value of abstract numbers, finally breaking down the archaic system of special numerations for various categories of items" (1981:330). These inscriptions allowed people to manage abstract quantitative information, and were rapidly proliferated to reference various types of objects.

The elaboration and refinement of numeric concepts coevolved with the expansion of temple-centered trade networks in the region. The earliest complex artifacts used to record amounts were recovered in the ruins of the major temple of Uruk, dated about 3350 B.C. Sumerian reliefs and drawings from that time display processions of individuals delivering goods to the temple (p. 31).

The societies that used numeric concepts to manage quantitative information and plan for the future were more likely to endure than the others, because of the applicability of numeric concepts. As Flegg puts it, "There is nothing in the physical world which is two. There are, however, a great many things in the physical world to which 'two' may be usefully applied" (1983:3). These numeric concepts obviously facilitated economic transactions. For example, in a society that had symbols only for one and two, people "would not willingly trade, say two cows for four pigs, yet had no hesitation in exchanging one cow for two pigs and a second cow for another two pigs" (Burton 1985:2).

Numeric concepts moved to a higher level of abstraction when group numbers came into use. Several different systems of group members were developed. Among the Egyptians a 1 (one) stood for a single unit, an inverted U stood for ten, and a spiral stood for a hundred; among the Maya a "." (dot) stood for a single unit; a "-" (bar) stood for five, and an elongated circle stood for twenty. Group numbers were developed to simplify recording procedures and were a cornerstone for the invention of computational procedures.

The conclusion that numeric concepts were first developed by western Asians may be based on the fact that the Sumerians inscribed in clay. However, many ancient groups used other materials to preserve quantitative information. The ancient Chinese used knotted cords and bamboo sticks to record quantity. Some ancients made notches on wood. Whereas clay tokens and tablets endure, wood, fibers, and bamboo sticks rot and leave no trace. But if numeric concepts were invented more than once, they seem to have always emerged from activity that centered on the storage of commodities. The numeric concepts of the Maya were originally developed in conjunction with trade.

The presence of a numeric system itself, even a rather complex one, did not assure that it would be applied to a wide range of phenomena. Members of a society with a lunar calendar could recite the names of the moons that compose the year, but may not have been able to specify how many moons there are in a year. Nor were they necessarily interested in the quantification of sequences if they had command of numeric concepts. The Tiv, for example, had a rather complex conceptual numeric system but did not use it for calendric purposes. If a Tiv were queried as to the number of moons in each year, the answers varied from ten to eighteen, and answers to queries of the number of days in a moon varied from ten to fifty (Hallpike 1979). Records of quantitative information existed along side of, but independent of, timekeeping procedures in many ancient civilizations.

Numeric concepts predate the emergence of monarchies and palaces. All ancient civilizations were using numeric concepts when they appeared on the historical scene. Furthermore, all ancient civilizations had specialists in the management of quantitative information. Some of the specialists were accountable to the ruling elite. All modern civilizations also have specialists in the manipulation of quantitative information who are accountable to the ruling elite.

Specialists in the management of quantitative information, such as accountants, recorded the amount of various goods on hand in ancient civilizations and they existed along side of, but independent of, the specialists in timekeeping. Much the same condition prevails in modern civilizations, where accountants manage quantitative information for ruling elites and have relatively little contact with scientists and engineers.

COMPUTATION OF QUANTITIES

The practice of inscribing numeric concepts on a surface did not lead to computational procedures. Instead, computational procedures evolved from using manipulative items to calculate changes in amounts on hand when a transaction was completed. As trade networks expanded and the amount of goods traded increased, computational procedures were developed. When a trader who already possessed seven sheep skins acquired six more, he could make six more incisions on a piece of wood to note the change. He thereby increased the amount recorded, but it would be stretching the concept of "addition" to include such incisions. Subtraction was probably the original computation. It is difficult to "erase" an inscription when the number of items on hand decreases. People began to subtract when they began to use matching objects to note changes in the amount of goods on hand, and the new amount was recorded.

Subtraction and addition are conceptual acts. When people use objects to subtract and add, they manage information by manipulating objects that stand for concepts. The objects manipulated do not reference concrete objects. The ancient computers used pebbles, sticks, and other small items to aid their manipulation of numeric concepts.

The word *calculate* is derived from the Roman word *calculi* which referred to any objects, most commonly pebbles, used on a flat area to add and subtract. The term *abacus* now refers to the combination of counters, columns, and frame that are used to compute. Its original meaning may have been "a flat surface" (Pullan 1968:89). The Chinese term for "computing device" is literally translated as reckoning board.

The pebbles, sticks, and bamboo shoots that were used to facilitate the most ancient computers have either disintegrated or are not easily recognized. Nonetheless, it is likely that the two sets of artifacts for managing quantitative information—incised notations and computing devices—coexisted for millennia. In all ancient civilizations for which there are historical records, pebbles and similar items were used to compute in conjunction with incised marks to record quantities. In addition, small stones and disks have been found at many levels of excavations in Mesopotamia that predate historical civilizations (Pullan 1968:8). It is likely that these artifacts were used to compute.

The ancient Egyptians used counting boards to add and subtract by counting forward and backward along rows of ten dots arranged in ten columns on a sheet of papyrus (p. 10). In a similar fashion, the Maya used grains of maize threaded in rows of ten much like the abacus in the Old World. A drawing on a vase found in Guatemala shows an Aztec official using an abacus (p. 93).

The Inca of South America used the quipu—a system of knotted cords—to record quantitative information. They also used counting boards. Several of the drawings of Inca record keepers show them holding elaborate sets of *quipu* (Zuidema 1983:56–66) and one of the drawings (p. 66) is of an Inca official holding a *quipu*, with a counting board displayed in the lower left-hand corner.

Computation by pebbles or an abacus seems complex to those not familiar with the procedure. It does require some training, just as learning to use a keyboard. Those schooled in the manipulation of an abacus, though, can proceed in much the same way as an expert typist does with a keyboard. They both calculate by touch.

PLACE VALUE

Several civilizations that developed group numbers did not develop the concept of place value. The ancient Chinese had the group number concepts of ten, one hundred, etc., but not the concept of place value. They recorded the quantity 462 by inscribing 4*x*, 6*y*, and 2*z*. Those notations indicated they were 4 hundreds, 6 tens, and 2 ones. But the ancient Chinese sometimes inscribed that quantity as 6*y*, 2*z*, and 4*x* and other times as 2*z*, 6*y*, and 4*x*. Such numeric systems facilitate the recording of quantitative information, but do not facilitate computation.

Place value systems were probably invented by specialists who managed information by simultaneously using two procedures for the management of numeric concepts. They made notches in pieces of wood or clay to record quantities and used pebbles or similar objects to compute changes in quantities. The development of place value concepts probably occurred as people began to use group numbers to record quantitative information and pebbles to calculate. For example, when they computed changes with pebbles, they may have used red pebbles to stand for one set of group numbers, gray pebbles for another group number that had a larger value, and so forth. If they used a group number system such as Roman numerals, they might have used pebbles of one color for one, another color for five, and still another color for ten.

Place value systems probably emerged as people began to arrange the pebbles representing different group numbers in rows as they computed. Once that practice became standardized, the locations of piles of pebbles with respect to one another could be assigned value. The pebbles on the left could be assigned the value of one hundred, those next to them the value of ten, and those on the right the value of one.

A place value system is a more efficient procedure for recording quantities than a system without place values, but the difference is not great. In contrast, it is very difficult to complete computations of any complexity without a place value system. Without place values, computers tend to become "lost in the confusion of a system of abstract written numerals" (Menninger 1969:371). Place values were part of the vocabulary "of all the peoples who developed or used counting boards" (p. 372).

Recording and computations procedures were not merged into a single system until relatively recently. Both technologies provide procedures for the management of quantitative information, but record keeping only requires command of numbers. Computations require both the command of numbers and procedures for the manipulation of numeric concepts. The merger of these two techniques was dependent on the development of a place value system for recording quantitative information as well as for computation procedures.

The Arabs accomplished that merger when they acquired awareness of a place value system after they invaded India. In Sanskrit, the original spoken symbol for zero was "empty." Empty meant there were no counters in a row on the counting board. The Arabic system numeric system has digits of 1 through 9. The difference between 3 and 5 in the Arabic system is the same as the difference between III and V in the Roman system. The Arabs combined their digital system with the symbol for empty (zero) and the location of a digit was assigned a value. Modern digital (base two) computers are the ultimate extension of the place value concept.

The Arabic system first entered western Europe as a record-keeping procedure. For several hundred years Europeans continued to compute with reckoning boards while using Arabic numbers to record quantities. The English royal treasurer continued to use tally sticks to compute until 1783 (Pullan 1968:51). That was more than two hundred years after "pen-reckoning" was in general use by merchants, bankers, and accountants (p. ix). Computations by counting boards were regarded as safer and more accurate than pen and paper reckoning. Also, pen and paper were valuable commodities that were not readily available to all who had occasion to compute.

STATES AND NUMBERS

About 3000 B.C., or perhaps a few hundred years earlier, a category of specialists in the management of quantitative information, known as *san-*

ga, emerged among the Sumerians. In the early period, the *sanga* were accountable to the *unken,* the city council of elders. Subsequently, monarchies emerged and the *sanga* became accountable to the monarchs, and egalitarian relationships were replaced with asymmetrical ones focused on the king. The emergence of the palace-centered state structures of the Sumerians was partially contingent on the merger of specialists in record keeping with specialists in violence. A similar evolutionary trajectory occurred in ancient China, Egypt, MesoAmerica, and South America.

"In every Inca settlement there were four official quipu keepers, known as amayocs, who tied the knots in these strings and submitted them to the central government" (Menninger 1969:25455). Only those trained in the procedure had access to the information. The Inca enlarged warehousing and other institutions as they established their empire, but a complex record-keeping system preceded the emergence of the Inca Empire (Collier 1982:12).

The Inca state was a conquest state and its continuation rested in part on coercion. But conquest has to be supplemented with a record-keeping system if state structures encasing large populations are to endure. The continuation of the Inca state was contingent on the accumulation, preservation, and secretive control of quantitative information by specialists. That information was used by state authorities to program state action and control citizens.

The Inca state was maintained by interlinking state elites with local administrators who supervised the granaries and communal labor forces (Wachtel 1982:199). When the Inca conquered a region the army was "succeeded by an Inca administrator, so that henceforth the economic function of the mitimas (the administrators) took precedence" (p. 202). Each community contributed foodstuff that supported the military and ruling elite, and the control of quantitative information was central to the establishment and maintenance of an autocratic state. Both the position of the ruling elite and that of the information specialists were rendered more secure by the monopoly of knowledge of quantitative information. That arrangement still prevails in varying degrees in many modern civilizations.

To maintain an egalitarian state structure, it is necessary for the citizens of the nation to have access to the quantitative information preserved by the record keepers of the state. Citizens of totalitarian states do not have access to such information, while to varying degrees citizens of representative democracies do. Only a few citizens of democratic states have the ability to decipher the complex records of modern societies. Nonetheless, when such information is public, those with the ability to decipher it can expose the misuse of communal resources. Many whistle-blowers lose their jobs when they make such information public, yet the threat of someone making information about corrupt use of funds public lessens the likelihood of corrupt state structures.

SPATIAL CONCEPTS

All societies order space, but some primitives could only do so with concrete symbols, because they did not have the concepts of point, line, square, circle, and so on. If shown a drawing of a circle and asked to specify its shape, these people might characterize it as being shaped like a dish. In societies without spatial concepts, people used such terms as "behind/in front," "above/below," and "right/left" to order space (Hallpike 1979:283). Even when people have command of spatial concepts they often use terms such as "left/right" and "next to" to specify spatial locations.

The belief that spatial concepts exist independent of human action and thought dates back to at least the time of Plato. But spatial concepts, like numeric ones, were invented by human beings as they confronted and were confronted by their material environment. And, as with numeric concepts, their development spanned millennia. The development of spatial concepts, like the development of numeric concepts, also was dependent on making markings to preserve information. The construction of diagrams to specify the relative location of concrete objects and events was a precursor to spatial concepts.

All ancient societies used points where the sun rises and sets to specify an easterly and westerly direction, and most primitive societies are aware of a spot in the heavens where the stars do not rotate and then use that spot to specify direction. The first horizontal locations used to specify direction with relative precision were where the sun rose and set at the solstices and where certain bright stars rose and set. When direction is specified by reference to a specific event (e.g., where the sun sets) the level of abstraction is the same as the specification of quantity with the terms *twins* and *triplets*.

All groups that used horizontal locations for timekeeping purposes came to recognize the necessity of making "standardized" observations. One southwest Indian group, for example, standardized their observations of sunsets by always making them "from the rooftop of the matriarchal house of the Bear Clan" (McCluskey 1977:197). After sedentary communities established a central observational position, they erected markers to specify the rising and setting of major stars as well as the risings and settings of the sun at the solstices and perhaps the equinoxes. Sticks and rocks were used to construct crude diagrams and to increase the precision of the observations.

The first spatial concept to be developed probably was the point, although it is possible that the concepts of point and line were formulated simultaneously. It may have been the case that the concept of point emerged only as lines were made that connected the points on the horizon to a centralized observational position.

Once a location on the horizon where a star or the sun rose was established, it "froze" the event created by celestial motion and transformed the event into an object. Originally the locations were probably referred to with such terms as "the winter's sun rising location" and "the winter's sun setting location." Much as the one-for-one markings were a necessary precursor to the formulation of numeric concepts, horizontal markings of locations were necessary precursors to the formulation of the concept of point.

When communities prospered and refined their monuments, they more precisely aligned parts of the monuments with specific celestial events. Many of these monuments, such as the so-called medicine wheels of the plains Indians, were composed of a series of lines of rocks radiating from a central viewing position to a series of locations on the horizon. Nearly all of the ancient diagrams that specified locations on the horizon of celestial events included lines that designated where the sun rose and set at the solstices. These sets of lines resulted in primitive diagrams. Diagrams are the foundation of geometry: "no diagrams, no geometrical way of thinking" (Unguru 1975:76).

Diagrams that depicted patterns of alignments created by the risings and settings of celestial bodies provided the foundation for the invention of spatial concepts. The most ancient engineering diagrams probably were developed by timekeepers who supervised the construction of communal monuments. Diagrams that display the relative positions of stars (celestial maps) may have been developed before terrestrial maps. The Sumerians and Egyptians had elaborate celestial maps prior to 3000 B.C., and those maps may have provided the foundation of the development of terrestrial maps.

The most ancient terrestrial maps that displayed the relative position of a number of locations and directions may have been developed by seafarers. Ancient seafarers invented port-to-port maps that provided information on how to proceed from port A to port B and from port B to port C, but those maps did not display area nor were they symmetrical. For example, the South Seas navigators had a system based on thirty-two concrete locations of the risings and settings of major stars. However, whereas the thirty-two directions of the modern mariner's compass are equidistant from one another, the directions used by the South Seas navigators were not. Their directional system was based on specific points and was neither conceptual nor symmetrical.

Grid maps that display area as well as direction probably were originally developed by agricultural communities in conjunction with the resolution of territorial disputes. Information specialists, such as the timekeepers affiliated with the temples of agricultural communities, may have constructed the first terrestrial maps.

The concept of line may not have emerged until after communities began building monuments that displayed shadows created by events such as the sun's rising at the winter solstice. The monuments of many ancient agricultural societies indicate that those who designed them had command of the concept of line. Several ancient Mesopotamian communities constructed temples that "conform with surprising consistency to a standard pattern of layout and decoration" by the sixth millennium before the Christian era (Oates and Oates 1976:13). The alignments of the ancient monuments of Britain indicate those who designed them had command of the concepts of point and line (Thom 1967, 1977).

Two-dimensional spatial concepts did not automatically flow from one-dimensional ones. Asymmetric grids were constructed before squares, rectangles, and triangles were, and ovals preceded circles. Between the seventh and fourth millennia before the Christian era, a number of temples were constructed in Mesopotamia, but the asymmetry of these buildings is striking (Aurenche 1986:76). The construction of these buildings seems to have been informed by the concept of line but not by the concept of square, rectangle, or triangle. Some of the buildings constructed during the latter part of this period appear to reflect symmetrical two-dimensional concepts. One temple built about 4900 B.C. contains at least two right angles (Seton 1984:42).

The civilizations of Central America had a complex grid system for organizing space but apparently did not have the concepts of square, rectangle, triangle, etc. Many ancient Central American monuments, temples, and state centers were laid out in an asymmetric grid that commonly deviated from a right angle by about 1-¼ degrees. Apparently specialists in monument construction developed sets of alignments keyed to specific points on the horizon and used the resulting grids to plot ceremonial centers and fields but did not develop two-dimensional concepts to structure those activities. Or, if they did develop two-dimensional concepts, they were different from the two-dimensional ones developed in the Old World.

Several gigantic drawings were etched in the desert of South America, and some of the mounds constructed by North American Indians display the configuration of animals. Two-dimensional spatial concepts appear to inform the construction of some of these mounds, but it is seems that symmetrical two-dimensional concepts were not widely used by those who designed them.

It is unlikely that symmetrical two-dimensional angular concepts—

square, rectangle, and triangle—emerged in conjunction with the construction of monuments designed to dramatize specific risings and settings of the sun and stars. In all probability those who designed these monuments were concerned about the proper alignment of their monument to celestial events and were indifferent to symmetry.

It is more likely that symmetrical two-dimensional angular concepts emerged from conflicts that centered on ownership of agriculture land. In primitive agricultural groups, the borders of fields usually follow natural markers such as streams, the edges of wooded areas, and hillsides. But when a region came under intensive cultivation, especially river valleys that flooded, conflicts over land ownership would come to the fore. The necessity of reestablishing field boundaries probably stimulated the development of symmetrical two-dimensional concepts. Some two-dimensional concepts were used in Mesopotamia by 3000 B.C., as reflected in some cities of the period. The city of Der had the form of a triangle, and the cities of Sur-Surrukin and Calah were laid out as squares (Oppenheim 1977:133).

By the Second Dynasty, about 2800 B.C., Egyptian monuments were informed by symmetrical two-dimensional concepts. But "[a] curious feature of Khasekhemui's tomb (constructed in the Second Dynasty) is its irregularity and faulty planning" (Emery as quoted in Hoffman 1979:350). In contrast, several monuments constructed by the Egyptians during the Third Dynasty, about 2700 B.C., take the form of a perfect rectangle.

A number of dramatic developments occurred in Egypt at this time in addition to the construction of gigantic pyramids. Heliopolis was established as an intellectual center and a written language was invented. It is conceivable that symmetrical two-dimensional angular concepts and perhaps circular ones issued forth from Heliopolis. Other developments of the same time included the invention of a numeric calendar and geometric formalization of sculpture and relief carvings. The intellectuals of Heliopolis may have been the first people to formulate symmetrical two-dimensional concepts. It may be that two-dimensional concepts were only invented once.

Thom (1967, 1971) and Thom and Thom (1978) offer evidence that the monuments of the ancient Britons were informed with the concept of circle. Some of these communities refined their monuments and specified a number of points on the horizon. They may have formulated two-dimensional concepts in conjunction with their monument construction activity.

The concepts of cube, pyramid, cylinder, cone, and sphere probably were first formulated as Egypt moved into the pyramid-building age. It is possible to build step pyramids such as those constructed by the Egyptians during the Third Dynasty without command of three-dimensional

spatial concepts. But the construction of true pyramids that do not collapse requires the presence of specialists who have command of the concept of pyramid and principles of analytic geometry. "Just as numbers
had to first be displayed before computations could be envisioned, so too
three dimensional symbols had to be displayed in artifactual form before
they could be used to design monuments" (Reese and Couch 1990:22).

Three-dimensional spatial concepts were probably formulated between the construction of the step pyramid in the Third Dynasty and the
construction of the Great Pyramid during the Fourth Dynasty. In the
intervening years, eight or nine gigantic pyramids were constructed. One
of the them collapsed when it "was given a novel, and yet untried, shape
by adding a mantle to transform it (from a step pyramid) into a true
pyramid" (Mendelssohn 1974:294). Refinements of three-dimensional
concepts and procedures for measuring and computing space probably
occurred as the intellectuals supervised the construction of the pyramids
and perfected their designs.

The Egyptian pyramids and the spatial concepts that informed their
construction were imbued with sacred qualities. For example, the three
sides of the Pythagorean triangle, which informed the construction of the
Egyptian pyramids, were given the names of three of their more important gods and goddesses. Osiris was equated with the three-unit side, Isis
with the four-unit side, and Horus with the hypotenuse (Seidenberg
1963). As was the case for the intellectuals of ancient Egypt, many modern
mathematicians regard geometric concepts with an awe that is absent in
their approach to numeric concepts.

SPATIAL COMPUTATIONS

The earliest computations were of quantities of items. The use of numeric concepts to compute length, area, and volume emerged subsequent
to the formation of spatial concepts and was refined as spatial concepts
were elaborated. The quantification of points was a simple achievement,
and the numeric concepts previously developed to quantify commodities
were used to note the number of points. The quantification of length was
a more complex achievement. The earliest quantification of distance was
in temporal terms. The *li*, a Chinese measure of distance, originally referred to a day's travel, and the *li* of rough terrain was shorter than the li
of smooth terrain. The *tupu* of the Inca, which was translated as equivalent to a league by the Spanish, "was longer or shorter according to the
terrain that had to be transversed" (Zuidema 1982:423).

Exactly how lines (length) were first quantified is somewhat problematic, but there are indications that the quantification of lines first occurred in conjunction with timekeeping and monument construction. The Sumerians used units derived from their measures of duration to quantify length. The Maya timekeeping system divided each day and night into nine units and one of their measures of length was nine units long (O'Brien and Christensen 1986:149).

The earliest quantification of the circle was also derived from timekeeping. The Chinese divided the circumference of a circle into 365-¼ units which is a derivative of the solar days in a year. The Egyptians divided the circle into 360 degrees about the same time as they adopted the 360 plus 5–day year. Both the ancient Egyptians and ancient Chinese used knowledge derived from celestial observations to quantify circular spatial concepts, but whereas the Chinese quantified the circle by applying asymmetrical units of measurement formulated on the basis of empirical observations, the Egyptians invented symmetrical units for the measurement of circular concepts.

The quantification of angular areas may have been achieved almost as early as the quantification of length. The Sumerians had quantified the square by 3000 B.C., if not earlier. "The oldest known (Sumerian) mathematical table is the table of large squares" (Friberg 1986:81). The Egyptians used symmetrical angular concepts to replot their fields and compute land areas in the Old Kingdom (Baer 1956). They may have first used the Pythagorean theorem to plot symmetrical fields and subsequently applied it in monument construction. The "manual of the cord" informed both the resurveying of fields and monument construction. The cord the Egyptians used for these undertakings consisted of a rope with twelve units which was divided into three, four, and five units to a side and thus simulated the Pythagorean theorem. The Pythagorean theorem is an extremely efficient procedure for laying out the foundations of buildings and specifying the boundaries of fields with ninety-degree corners. It continues to be used for that purpose.

The civilizations of the New World, such as the Maya and Inca, may have quantified area, but if they did, they apparently did it without benefit of symmetrical two-dimensional concepts. The Inca who used the term *tupu* to reference distance also used that term to refer to area, but the term did not refer to a quantified area. Instead it "corresponded to the land needed to support a family" (Zuidema 1982:423).

The Egyptians of the Old Kingdom developed elaborate and precise procedures for quantifying area. For example, they developed a standardized procedure for doubling the area of a square. The three basic units of the Egyptians for quantifying length were the *remen*, the *cubit*, and double *remen*. "The double remen was the diagonal of a square whose side was 1 cubit." (Lumpkin 1980:186). By changing the unit of measurement from a *cubit* to a double *remen*, the area of a square would be doubled. By changing from a *cubit* to a *remen* the area would be halved. The procedure produces "irrational" units of lengths but is an effective procedure for managing two-dimensional space.

The Egyptians apparently were the first to develop systematic procedures for the computation of volume and surface area of three-dimensional forms. One was a formula for computing the volume of a truncated pyramid (Gillings 1972). One indication that the Egyptians were the first to compute the volume of three-dimensional forms is that the most ancient procedure known for computing the volume of a cube in the Mesopotamia region used the unit *mindan* for the horizontal dimensions of the cube and the term *cubit* for the vertical dimension. It is likely that the measure *cubit* and the procedure for computing the volume of a cube used in ancient Mesopotamia were imported from Egypt.

The development of symmetrical two- and three-dimensional concepts and their qualification appears to have first occurred in Egypt after or simultaneously with the emergence of a written language and the establishment of Heliopolis as an intellectual center. The Sumerians were literate before the Egyptians, but the Sumerians wrote on clay whereas the Egyptians perfected the manufacturing of papyrus. Papyrus provided the Egyptians with a more utilitarian surface than clay for drawing diagrams, making notations, and calculating. Another factor that may have contributed to the innovativeness of the Egyptians during the early dynasties was that the Egyptian nation contained two distinct cultural traditions. That may have created a context more conducive to innovations than that which prevailed in the Mesopotamian cities.

CONCLUSIONS

The development of numbers that were not content bonded created the first set of abstract mathematic symbols (concepts). The invention of group numbers brought about a second level of abstraction. The invention of place value created a third level of abstraction. Conceptual thought, as opposed to concrete symbolic thought, became more elaborate as people mastered the manipulation of artifactually preserved information about amounts and spatial locations. The development of these concepts allowed people to engage in "pure thought" or logic. These concepts referenced no specific objects, but could be used to organize many different kinds of experiences and actions.

The development of procedures for visually presenting concepts provided a foundation for the subsequent development of complex bodies of abstract knowledge. The word *idea* is a derivative of a Greek word meaning "the appearance of a thing" and the word *theory* is a derivative of a Greek word meaning "to view" (Logan 1986:121).

The emergence of numeric concepts was coterminous with the emergence of markets, trade networks, and the storage of large quantities of goods, especially hard cereals. Systems of computation probably did not emerge until after sedentary communities with populations of at least a few hundred and perhaps a few thousand had been established. Numeric concepts and computational procedures emerged out of repetitive social endeavors and gave greater stability to the social endeavors from which they emerged. The expansion of these trade networks and the associated growth in populations of the sedentary communities were contingent on the elaboration and refinement of procedures of quantification. The development of numeric concepts and computation procedures in turn allowed for the subsequent establishment of larger communities and ultimately state structures. Specialists in the management of quantitative information are focal persons in all state structures. In totalitarian states, quantitative information about state resources usually does not circulate beyond the confines of state officials.

Some state structures were established and endured without literacy, but all of them have specialists in the use of numeric concepts. The emergence of state structures was not contingent on all citizens acquiring mastery of numeric concepts and computational procedures. Just as only a small percentage of the citizens of the United States is capable of the computations necessary for the installation of a computerized bookkeeping system, only a few savants of ancient societies were capable of complex computations.

Despite the fact that only a small number of the citizens of ancient civilizations had the ability to compute on a conceptual level, it transformed the lives of all. Today the installation of a computer system by the U.S. Internal Revenue Service has modified the relationship between citizens and the state: citizens have become more accountable to state officials.

The conciseness, precision, and abstract nature of numeric concepts allow human beings to undertake social endeavors and create social arrangements not otherwise possible. State record keeping, taxation, international trade, highway construction, airplane travel, and welfare systems would not be possible without numeric concepts and computation procedures.

Spatial concepts were elaborated and refined as societies built monuments to display celestial events and the boundaries of fields were plotted and replotted. When spatial concepts were merged with numeric ones, analytic geometry was created. The division of the Earth into degrees of latitude and longitude to facilitate long-distance travel, the use of 90-degree angles to lay out foundations of buildings, and the 360-degree compass to specify direction are but three examples of how spatial concepts inform how human beings experience and act toward their material environment.

Analytic geometry allows human beings to undertake elaborate social enterprises that rearrange the material environment. Principles of analytic geometry inform such varied activity as building skyscrapers, laying sewage lines, and stationing outer-space probes. Analytic geometry, more than any other information technology, provides a means by which human beings impose symmetrical order on their environment. Edifices ranging from the Great Pyramid to the lawns of modern suburbanites reflect geometric concepts.

Many ancient civilizations used spatial concepts originally formulated to abstractly characterize celestial events and objects to structure some social relationships. The Inca Empire was divided into forty-two divisions with each division attached to one of forty-two sightlines that radiated "from the central temple of the Sun to the horizon" (Zuidema 1983:54). A noble was in charge of each sightline and the settlements adjacent to each sightline. The Chinese used principles derived from their astronomic studies to structure their state bureaucracy: "The names of the constellations and those of the bureau of the imperial administration corresponded" (Yabuuti 1973:94).

Geometric concepts were one of the necessary conditions for the emergence of science. They imply that a perfect symmetrical order exists. That belief continues to inform research dedicated to the formulation of universal principles. The failure of the Chinese to develop a scientific mode of thought may have been due to the absence of a set of symmetrical spatial concepts that were integrated with numeric concepts. "Chinese astronomy remained basically an algebraic celestial kinematics of the Babylonian type for more than two Millennia" (Nakayama 1973:34).

Mathematic concepts and procedures allow for the programming and completion of social endeavors that are beyond the imaginations of those who do not have command of them. They thereby have provided opportunities. But in many instances, especially when shrouded with secrecy and mysticism, these technologies have also enhanced the ability of those with command of them to exploit others.

CHAPTER

6

Bureaucratic Structures

Bureaucracies are information technologies in the same sense that assembly lines are material technologies. Both are social arrangements that endure despite changes in personnel and use standardized procedures to achieve collective objectives. The collective objective that informs the actions of factory workers is the production of goods; the collective objective that informs the actions of bureaucrats is the management of information. Bureaucracies are second-order technologies that use first-order information technologies (numeric concepts, calendars, written languages, computers) to manage information.

Bureaucratic structures emerge out of the record-keeping activity of an encompassing social system. An extended family acquires bureaucratic elements when one family member has the responsibility of keeping a record of one type of family resource, a second another resource, and both are subordinate to a third member. Just as record keeping is always an adjunct of other activities, bureaucratic structures are always adjuncts of larger social structures. Each bureaucracy accumulates and preserves information for an organization. No bureaucracy is an autonomous social unit. Bureaucrats are not authorized to direct the activity of the organization they service but they often influence the actions of the encompassing organization by influencing the flow of information.

In mature bureaucratic structures each person has specific responsibilities, is accountable to a specific other, and has a rank in a hierarchical order. Some bureaucrats accumulate information, others preserve it, others dispense it, and still others supervise those who manage it. The primary objective of bureaucratic structures is to provide information to their encompassing organization so that the organization can more effectively accomplish its objectives.

Each bureaucrat, including the head bureaucrat, is obliged to provide specific types of information to specified others. The other may be the chief executive of a company, a committee, or an on-line member of an organization. Bureaucrats only have internal authority; they do not have authority to issue directives to nonbureaucrats. Chief bureaucrats of state bureaucracies have become heads of state, but when that occurs their primary responsibility is no longer to manage information, but to program and supervise the actions of an organization.

The distinction between bureaucrats and nonbureaucrats, like all dichotomies, is sometimes ambiguous. In large factories, accountants manage information; on-the-floor supervisors use the information to supervise assembly line workers; and those who work on the assembly line may use information provided by bureaucrats to structure their actions. Information from the assembly line workers is accumulated and transmitted to the record keepers, who organize and preserve it. Grammar school teachers are obliged to take class attendance and pass the information on to a record keeper. However, the primary responsibility of teachers is to teach and career advancement supposedly is based on their teaching. In contrast, career advancement for bureaucrats is based on their effectiveness in managing information.

There is great variation from organization to organization in the degree of bureaucratization. In some, the division of labor between managers of information and others is not clearly delineated, while in others the distinction between record keepers and others is rigidly drawn. Fairly complex social structures have been created without the benefit of clearly delineating record-keeping activity from other types of activity, but all modern states, industrial complexes, and cities contain a bureaucratic structure.

Some scholars have equated the formation of bureaucratic structures with the emergence of civilizations (Wright 1977; Wright and Johnson 1975). When civilizations emerged in Mesopotamia and the Nile Valley they contained bureaucracies, but several complex ancient societies emerged without benefit of a bureaucracy. The ancient Britons constructed Stonehenge and many other megalithic monuments, and there is no evidence that they were blessed with a bureaucratic structure. As a result, students of ancient societies have been reluctant to classify societies such as the ancient Britons as civilized.

The across-generation continuation of complex social structures seems to be contingent on the development of bureaucratic structures, but the emergence of bureaucratic structures does not assure the continuation of complex structures. Some fairly complex structures have been constructed without benefit of bureaucratic structures, but there seems to be a level of complexity that organizations cannot surpass without benefit of bureaucratic structures.

EARLY BUREAUCRACIES

The first steps toward bureaucratic structures may have been taken in western Asia as large communities prepared for intervillage feasts in which numeric tokens were used to record the "pledges of food offerings to be delivered by individuals and groups at the time of ritual ceremonies. . . . The tokens permitted the estimation and computation of the total budget of goods which could be depended upon" (Schmandt-Besserat 1982:876). This information probably was accumulated by the specialists responsible for organizing the festivals. Their management of information about the goods provided by other villages was probably subsidiary to planning the festivals: it is doubtful that the primary obligation of these specialists was to keep records.

The most ancient bureaucracies were appendages of temple-centered structures that were constructed as temple-centered trade networks expanded in Mesopotamia in the latter part of the fourth millennium. Much of the intercommunity trade of the Sumerians was funneled through the temples. The procedures for recording the goods stored at and traded by the temples were elaborated as trade networks were enlarged. The record-keeping procedures for the temple-centered trade included the use of tokens by the *sanga,* which is often translated as priest, but a more accurate translation is accountant or record keeper (Biggs 1974).

In the early period the *sanga* were not ranked, but as the volume of trade increased the number of *sanga* increased and rankings were established. The *sanga* lived among other residents in the early period, but later their living quarters were attached to the temples (Adams 1966:135), which suggests that a degree of social distance emerged between the *sanga* and other members of the community as the Sumerian bureaucracy matured.

The early temple-affiliated bureaucrats of Mesopotamia were accountable to the *unken,* or a council composed of elders. Later, as palaces emerged and competed with the temples for communal dominance, bureaucrats who were affiliated with the palaces appeared. The procedures of record keeping and the ranking of the record keepers remained relatively constant for centuries. There is "an unbroken text tradition spanning more than seven hundred years" from the earliest bureaucracies to the later palace-affiliated bureaucracies (Nissen 1988:80). These record-keeping procedures diffused from central Mesopotamia "in all directions in the Near East and was connected in part with the setting up of 'colonies'" (p. 107).

Bureaucratic structures emerged in Egypt a little later. The earliest Egyptian record keepers may have been affiliated with the palace. From the very earliest period of the Egyptian nation, there was a chief bureaucrat, the grand vizier, who was accountable to the king. Statues of scribes became common in the Fourth Dynasty which suggests that record keepers occupied a central position in the state structure at that time. In the early dynasties the rankings of Egyptian bureaucrats varied from king to king, but a definitive ranking of the Egyptian bureaucrats was established

in the Fifth Dynasty (Baer 1956). Subsequently, the Egyptian nation became one of the most bureaucratized societies of all time.

Centralized record keeping developed at least a few centuries before
palace-centered state structures emerged in Mesopotamia, and may have
developed simultaneously with the monarchy in Egypt. Complex record-
keeping procedures were in use for at least a few centuries before the
arise of the Inca Empire. These data indicate that bureaucratic structures
are precursors of empires, rather than a consequence of them.

The expansion of ancient palace-centered states included the expansion of bureaucratic structures. The enlargement of the Chinese state, for
example, was coterminous with the establishment of local bureaucracies
accountable to the chief bureaucrat of China following the conquest of a
region. In China, a major transformation "began, perhaps, in 213 B.C. with
the destruction of local records by the first emperor" (Elvin 1973:21–22). A
similar practice was followed by the Inca. Each time the Inca army conquered a new region, they were followed by state bureaucrats who recorded information about local resources and transmitted the information
to the capital city. The Romans also imposed their bureaucratic structure
on those they conquered.

The growth of bureaucratic structures was accompanied by the establishment of scribal schools attached to temples or palaces. Those trained at these
schools were relieved of manual labor and obtained the necessities of life
from tithes and taxes. The faculty of these schools taught the students not
only the principles of record keeping but the rationale for their elite position. The Chinese bureaucrats offered this rationale: "Some labor with their
minds, some labor with their strength. Those who labor with their minds
govern others; those who labor with their strength are governed by others.
Those who are governed by others support them; those who govern others
are supported by them" (cited in Balazs 1964:62).

As bureaucratic structures became a significant feature of ancient palace-centered states, bureaucrats began to wear apparel to distinguish
themselves from others. The ancient Chinese bureaucrats grew long fin-
gernails, which announced to all that they did not do manual labor.
Statues of scribes from the middle kingdom of Egypt depict them with
rolls of fat on their ribs. The modern distinction between white- and blue-
collar workers is a muted form of the same distinction.

The early Sumerian bureaucrats maintained a running account of resources stored at the temples. One of the earliest records kept in Egypt
was the height of the yearly floods, which was used to establish taxation
rates. Taxes were in kind: farmers were obliged to contribute a percentage
of their crops each year to the state. By the Second Dynasty the Egyptian
bureaucrats were taking a census of people, livestock, and fields every
two years.

The communities of ancient Mesopotamia contained "big families" who maintained continuity across generations in record-keeping procedures "despite changes in politics and government" (Gibson and Biggs 1991:3). On several occasions, families of bureaucrats competed with warrior elites for dominance of the state structures of ancient societies. For example, in ancient Egypt during the First and Second Dynasty the head bureaucrat, the grand vizier, was a relative of the king, but by the end of the Fourth Dynasty the bureaucrats owed their allegiance to the state bureaucracy instead of the ruling lineage (Baines 1988:203). The Fifth Dynasty was instituted by the bureaucratic elite supplanting the royal lineage. That was but the first of several transformations in Egypt wherein bureaucrats acquired control of the state structure. On several occasions the bureaucratic elites of state structures have simultaneously served heads of state and plotted their overthrow. Conversely, on several occasions heads of state have lost their patience with their bureaucrats and attempted to teach them their proper position in state structures. In 213 B.C. the emperor of China beheaded a number of bureaucrats.

The first bureaucracies, just as with other information technologies, were associated with particular activities and were developed to enhance those activities. The Sumerian *sanga* were not created to establish a new structure, but to facilitate intercommunity trade. Nonetheless, those early record keepers laid the foundation for the emergence of a distinctive set of social relationships—relationships that exist solely for the purpose of processing information about communal resources.

Bureaucratic structures create monopolies of knowledge that empower bureaucrats. But those monopolies of knowledge do not provide all the resources that are necessary for bureaucrats to effectively control others. To accomplish that, it is necessary for bureaucrats to form coalitions with others, especially those who control coercive forces. Bureaucrats have often formed a coalition with warriors to exploit rank-and-file citizens. For example, resistance to payment of taxes in ancient Egypt was minimized by having a soldier accompany each tax collector. Modern tax collection systems follow much the same procedure, although the threat of coercion is usually more subtle.

THE FORMATTING OF INFORMATION

The cliché "Everyone is just a number" is a caricature of bureaucratic mentality, yet reflects a central feature of bureaucratic activity. The primary obligations of bureaucrats are to identify and to quantify. Until recently, all objects and people recorded were named, but in modern bureaucracies the identities of objects and people are also rendered in numeric terms. Each citizen of the United States is required to have a social security number. The quantification of everything depletes a concern with nuances and variety.

Unlike those who populate most other social structures, bureaucrats do not, except incidentally, confront the material environment as they complete their tasks. Instead, they deal with concepts that reference other objects. For the most part bureaucrats process disembodied referential concepts. The information does not, except for that which specifies identities (and not even that when numbers are used to specify identities), reference variation in quality.

Bureaucratic action in its pure form is divorced of all evocative symbols. All, including self, are but numbers. But all are not merely numbers, because not even the completely bureaucratized person organizes his actions solely on the basis of numeric concepts. Each bureaucrat has hopes and fears. Even if he has lost hope for advancement, he hopes to retain his position. He knows others have little compassion for his hopes and fears, that they attend to him almost solely in terms of his utility to them. Experienced bureaucrats are woefully aware that they are merely instruments to be used by others. Authorities select and discard bureaucrats much as automobile owners select and discard automobiles.

Bureaucrats are not the only people who routinely make extensive use of numeric concepts. Engineers and scientists also do so, but engineers and scientists test their concepts by using them to confront an external environment. On occasion, bureaucrats also must deal with an external environment, but more often the demonstration of the validity of their numeric concepts is achieved by noting if they can be aligned with established categories and procedures.

The procedures developed by bureaucrats to facilitate the processing of information usually reflect the interests of bureaucrats more than the interests of clients. The formats adopted inform encounters between bureaucrats and nonbureaucrats and among bureaucrats themselves. The impact of bureaucratic procedure often is subtle but pervasive.

The objective that originally informs the establishment of standardized procedures is to enhance communal welfare but that objective often does not inform assessments of a bureaucracy's operation. Instead, assessments of bureaucratic activity are commonly made, especially by bureaucrats themselves, on the basis of whether or not the information is properly formatted.

EXTERNAL RELATIONSHIPS

The most significant external relationship for each bureaucracy is that of the chief bureaucrat to the authority of the encompassing organization.

The grand vizier dispensed information to the Pharaoh. In turn, the Egyptian tax collectors acted in behalf of the Pharaoh, and citizens were required to abide by the dictates of the Pharaoh as interpreted by the bureaucrats. In a similar manner, it is the interpretations of local agents of the Internal Revenue Service that confront each taxpaying U.S. citizen. There has been a close interweave between bureaucratic structures and control agencies from the institutionalization of the "eyes and ears" of the Pharaoh in Egypt to the FBI in the United States.

All bureaucrats are aware that their personal welfare is contingent on pleasing their superordinates, and so encounters with external superordinates are approached with caution. In contrast, when bureaucrats are attempting to implement directives by acting as delegates of their authorities, their encounters with clients often are adversarial and they usually are able to maintain control of the tenor of encounters with clients. They usually have command of more relevant information than clients and that usually allows them to impose bureaucratic formats on most of their clients. Accordingly, most adversarial encounters between bureaucrats and clients are resolved to the bureaucrats' advantage.

Despite the fact that bureaucrats are subordinate to an external social structure, many have been able to transform that relationship to their advantage. Chief bureaucrats of state bureaucracies have commonly used the information at their command to transform their relationship with heads of state into an egalitarian symbiotic one. For example, J. Edgar Hoover, when head of the Federal Bureau of Investigation, accumulated profiles of U.S. presidents and informed each of them that he had command of information that would embarrass them if he made it public. Many of those supposedly serviced by bureaucrats have found themselves confronted by an obdurate reality constructed by bureaucrats and dependent on their record keepers.

Cynicism and corruption is a common consequence of bureaucratic structures, and bureaucrats often become alienated from and come to think themselves as superior to those they serve. Reciprocally, many nonbureaucrats become frustrated with and alienated from bureaucrats. Bureaucrats elicit little sympathy from outsiders who, in turn, have little sympathy for them. Solidarity is seldom generated between bureaucrats and those they serve.

INTERNAL RELATIONSHIPS

The actions of bureaucrats are similar to those of assembly line workers in that work goals and means are specified and each person acts almost as an automaton. In the ideal bureaucracy, each person acts toward information in isolation from other members of the bureaucracy. Other bureaucrats may be present, but bureaucrats seldom act in unison to achieve social objectives. Accountability, therefore, rather than solidarity pervades bureaucratic structures, and bureaucratic accountability is based primarily on products, not on activity. That is, it is bureaucratic records that count, not actions. Many bureaucrats give primary consideration to the information they process and only secondary attention to whether or not the original objectives that inform the encompassing organization are achieved. Whether or not a million bushels of wheat were in fact produced often becomes less important than that the records indicate that a million bushels were produced.

Bureaucratic activity has been characterized as the epitome of rationality. If rationality is defined as taking action in the present to obtain an objective, however, then bureaucratic action can be more accurately characterized as nonrational. It is common for those producing records, for instance, to assess them in terms of whether they are consistent with established formats. Whether or not the records contribute to movement toward the objectives of the encompassing organization often is irrelevant to assessments of record-keeping activity.

Many bureaucratic administrators use established formats to assess the performance of lower level bureaucrats. Consequently, the accomplishments of subordinates often come to be less significant than whether or not proper form was followed. Low-level bureaucrats seldom innovate or initiate. Rather, they follow procedures, and often provide information that is merely reflective of their superordinates' wishes. In extreme instances, the information provided has so little validity that the encompassing organization collapses. The ideal bureaucrat, as assessed by his bureaucratic superior, mechanically processes information by applying a specified format to the information processed.

Bureaucrats usually are concerned the most with anticipated assessments from their superordinates rather than the assessments of peers. Consequently, bureaucrats seldom develop robust solidary relationships with one another. Compassion for one's fellow record keepers usually is subordinated to self-interest, and egoism prevails. Egoism and the relative absence of solidarity within bureaucracies inhibits the likelihood of collective protest by bureaucrats. Only rarely have bureaucrats mounted social protests, and they seldom go on strike. Each record keeper advances by competing with fellow bureaucrats for recognition from superordinates.

Despite the fact that bureaucratic structures generate a concern with assessments elicited from superordinates, they do not generate loyalty to superordinates. Bureaucratic subordinates seldom commit altruistic acts for their superiors. When they cater to their superordinates it usually is a calculated act of deference, which contributes to a culture of cynicism and egoism rather than faith and altruism.

Those who supervise other bureaucrats frequently attempt to control the internal flow of information as well as the flow of information from the bureaucratic members of an organization to external social units. Although the lower-ranking members of a bureaucracy have little opportunity to control the flow of information, all but the most naive newcomers are aware that the flow of information is typically controlled by those capable of doing so, which also generates cynicism, especially among the lower ranking bureaucrats. Each becomes primarily concerned with personal well-being. In Weber's terms, the primary interest of most bureaucrats is how to become a bigger cog in the machine.

Once a bureaucracy is in place, bureaucrats usually influence the programs of action formulated and sometimes are able to have policies formulated that resound to the benefit of bureaucrats. The adeptness of bureaucrats in managing information often allows them to siphon off communal wealth. Corruption is often rampant. When bureaucrats can keep nonbureaucrat outsiders uninformed about communal resources, they increase their opportunities for personal gain.

That bureaucratic structures generate a distinctive morality is revealed by the fact that from the time of the ancient Sumerian *sanga* to the present day, most bureaucrats have willingly served conquerors. When the Mongols conquered China, they left the Chinese bureaucracy intact; when the Allied forces conquered Germany in 1945, they left the German bureaucracy intact. In these and hundreds of other instances, the bureaucrats served their new masters as efficiently as they had served their old ones. In the words of one, morality in a bureaucracy "is not what is right in a man's home or in his church. What is right is what the guy above you wants from you" (as quoted in Jackall 1987:90).

Some bureaucracies, in combination with coercive forces, have become so efficient in exploiting those they supposedly serve that the structures they served have collapsed. One such instance was the Roman Empire. The state structure of the Roman Empire in the early centuries was dominated by a warrior elite, while in the later centuries it was dominated by a bureaucratic elite. The high degree of bureaucratization of the Roman Empire is signified by the fact that some of the later Roman emperors were career bureaucrats.

IN THE SERVICE OF EVIL

Bureaucratic structures have generated corruption, callousness, and cynicism from at least the time of the Fifth Dynasty of Egypt. The epitome of the evils associated in bureaucratic structures was achieved by the state

bureaucracy of Germany during the Nazi regime. That bureaucratic structure was encased by a totalitarian state that rested upon a vibrant charismatic foundation (Couch 1989). The state structure created by the Nazi
party was infused with a fiery charismatic relationship that was focused
on Hitler, a warrior class (the storm troopers), and secret police (the
Gestapo). These factors gave the state structure of the Nazis a distinctive
character, but Nazi Germany could not have achieved the epitome of evil
that it did if an effective bureaucratic structure had not been available to
be harnessed by the Nazi party. A highly bureaucratized state structure
was in place long before the Nazis acquired control of the state apparatus
(Hilberg 1961).

Hilberg advances the argument that the destruction of the European
Jews was a consequence of the bureaucratic structure of the German state.
The state bureaucracy of Germany contributed to the achievement of that
objective by managing the information that informed the arrest, transportation, and murdering of Jewish citizens, but the impetus for the destruction of the European Jews did not originate from within the German
bureaucracy. It was the Nazi party, under Hitler's charismatic leadership,
that mobilized anti-Semitic sentiment and formulated the social objective
of destroying the Jews. The achievement of that objective required using the state bureaucracy to implement the policies formulated by the
Nazi party.

The German bureaucrats transformed the policy formulated by the
Nazi party and put it into practice. Very few bureaucrats required prod-
ding; most willingly took on their assigned tasks (Hilberg 1961:19). When
the policy of the so-called final solution was instituted, most of the bureaucrats approached the assignment as merely another function to perform. As Hilberg puts it, "The machinery of destruction was structurally
no different from organized German society as a whole; the difference
was only one of function" (p. 640).

Tens if not hundreds of thousands of Jews were massacred before
murdering them became bureaucratized. The massacres were encouraged
by the state elite but they were not done systematically. Warrior elites, the
storm troopers, carried out several early massacres. However, many professional military men regarded killing unarmed people as beneath the
dignity of warriors and resisted assignments to kill Jews. Furthermore,
Nazi leaders became concerned about the psychic burden of the soldiers
called on to murder unarmed Jews (p. 218).

The leaders of the Nazi state therefore turned to the state bureaucracy
for a solution. The program for the final solution was "drawn up by very
simple people" (p. 19) and implemented by the state bureaucracy. The
leaders of the Nazi party attempted to distance themselves from the killing
yet attempted to control how the killing was to proceed. A Nazi party
directive asserted, "It would be beneath the dignity of the 'movement' if its

members were to molest individual Jews" and the amateurish handling of the problem "was to be avoided under all circumstances" (p. 29).

In 1943, as the procedures for murdering the Jews and other "undesirables" were perfected, a bureaucratic memorandum noted, "[W]e have no right to take a single fur, a single watch, a single mark, a single cigarette, or anything whatever" so that in the end we can say "we have not been damaged in the innermost of our being, our soul, our character" (p. 647). Several edicts called for those killing the Jews to take a detached attitude in the performance of their tasks. The edicts called on those who organized the murdering not "to look to the right or to the left" (ibid.). The detached and narrow focus adopted by many bureaucrats rendered their evil activity banal, and many approached their task as merely another bureaucratic obligation.

Many edicts were issued stating that the killing was to follow standardized procedures. If an unauthorized killing "was an expression of idealism, no punishment was necessary unless the maintenance of order required disciplinary action or prosecution. However, if selfish, sadistic, or sexual motives were found, punishment was to be imposed for murder or for manslaughter" (p. 648).

"The bureaucrats were fully aware of the dangers of plundering, torture, orgies, and atrocities" (pp. 646–47), and they fretted about excesses such actions might generate among those who had to personally confront their victims. One procedure developed to minimize expressions of sadism was "to charge inmates instead of guards with beating the prisoners" (p. 578). "The bureaucracy was not so much concerned with the suffering of the victims as with the contamination of the perpetrators" (p. 577). When an evil activity is bureaucratized, so it seems, the evil inflicted creates as much suffering for the victims as nonbureaucraticized evil, but the actions become banal from the point of view of the perpetrators, and thus is just one more task to accomplish.

While some Nazi bureaucrats spoke proudly of improvements in the killing processes they had perfected (p. 572), others aided the escape of a few of those destined for murder. At the war crime trials, most of the defendants produced evidence that they had helped some Jews. When the bureaucrats adopted a bureaucratic attitude, they appeared to have had no difficulty in sending their fellow citizens to their death, but when they adopted a community- or family-based standpoint some acted with compassion. The humane acts of off-duty bureaucrats, however, were petty in comparison to the evil they inflicted when on duty. Yet "[t]hey separated 'duty' from personal feelings. They attempted to preserve a sense of 'decency'" (p. 659).

The bureaucrats "realized the connections between their paperwork and the heaps of corpses" (p. 658), but rationalized their actions as a necessary evil. They were good people doing dirty work. The German bureaucrats were aware that another viewpoint, other than a bureaucratic one embedded in the ideology of the Nazi party, could be taken when assessing their actions. When the bureaucrats recognized the impending defeat of Germany a few committed suicide. So long as the Nazi party retained control of the state apparatus though, the accountability of the bureaucrats to the Nazi elite suppressed the likelihood of them adopting an alternative standpoint toward their actions.

The bureaucracy of the German state in the service of the Nazi party is an extreme case. Nonetheless it demonstrates a feature common to all bureaucratic structures, namely, a focus on procedures, the adoption of the standpoint of one's superordinate toward one's own actions, and a lack of concern with the consequences of one's action for others.

The German state was not the only one to institute wholesale slaughter. The ancient Romans on several occasions annihilated populations. The distinctive features of the German accomplishment was that it was enduring, not sporadic, and followed standardized procedures. It was performed largely as a duty, not as an expression of personal or collective anger. The long-range consequence of bureaucratic evil is to render people largely devoid of compassion.

BUREAUCRACIES AND SOCIAL SCIENCE

Some social theorists have advanced the argument that with the growth of bureaucratic structures a particular type of social system—the meritocracy—has become dominant. That frame of thought presumes that prestige and rewards are acquired on the basis of personal merit instead of lineage, success in the market, communal service, or success in war. What is generally not recognized is that the meritocracy is based on a particular set of criteria that are formulated by bureaucrats. These criteria include the ability to manage quantitative information, to abide by directives from superordinates, to accept established procedures, and to avoid compassion. Such criteria, of course, reflect particular sentiments and interests as surely as those that are based on prowess in war, success in the market, artistic fame, communal service, or compassion.

Most of the research on bureaucracies presumes that they are necessary and asks the questions: How do they work? How can they be made more efficient? The theories of bureaucracy advanced are often little more than disguised rationalizations for bureaucratic structures (Simon 1957; Padgett 1980). The possibility that other social relationships might be constructed to manage information about communal resources is seldom advanced by social scientists. Instead, "[t]he bureaucrat's version of how bureaucracy works has been pressed into the service of science by modern functionalists to explain not only bureaucracy but social action in general" (Hilbert 1987:71). The concepts of role, role set, role expectations, status, values, and norms have been used by functionalists to formulate theories of human conduct. These concepts presume that the mechanistic paradigm is adequate for study of social processes. Such concepts may be adequate for the analysis of bureaucratic activity, but they are far from adequate for the analysis of most social processes and relationships.

The functionalist/bureaucratic paradigm also presumes that quantitative information, social facts, are the proper objects of investigation for social scientists. Emile Durkheim, the founder of sociological functionalism, claimed that it was not possible to observe social processes and relationships, and therefore social products, social facts accumulated by bureaucrats, were the proper objects of observation and analysis. That paradigm has spawned innumerable research projects that analyze the incongruence of action with bureaucratic norms. Nearly all such analyses are based on the presumption that actions not congruent with bureaucratic norms are pathological.

When social scientists use the bureaucratic paradigm to conduct research on social structures, they often use the definitions of reality offered by bureaucratic authorities to specify the objectives and procedures of their research. That practice presumes that high-ranking bureaucrats have wisdom.

The bureaucratic paradigm implicitly denies that human action is willful and emergent. But social structures are created and maintained by people acting with intentionality and even the most rigid social structures constantly evolve and devolve. The establishment of a bureaucracy clearly is a willful act. As bureaucratic structures mature, the social objectives that informed their creation often recede and are supplanted by willful acts undertaken to promote personal advancement. However, the actions that maintain bureaucratic structures also are willful. All social action is given structure by persons projecting futures and organizing themselves to bring a projected future to fruition. The actions of no one, not even those of the most bureaucratized person, merely reflect norms and values.

All bureaucrats recognize that not all action can proceed according to established formats. Yet the continuation of each bureaucratic structure is partially contingent on general acceptance of that belief. One irony is the belief among many social scientists that the paradigm of thought implied by bureaucratic structures is adequate for the analysis of social life. Only the most naive bureaucrats think that formal procedures are adequate for all situations; experienced bureaucrats know human action does not confine itself to bureaucratic formats. It would seem that specialists in the study of social structures would be as sophisticated about social structures as bureaucrats.

Bureaucratic structures are one procedure for the management of large bodies of quantitative information. But bureaucratic structures are not the only procedure for processing information about communal welfare, and they certainly are not the ultimate social form. Instead of conceptualizing them as the epitome of rationality as functionalists do, one might conceptualize them as an aberrant social form. A question that seems worthy of attention is: How can large bodies of information be managed without the creation of a set of social relationships that generates the sentiments and interests peculiar to bureaucracies?

CONCLUSIONS

Bureaucratic structures have evolved from their first appearance five to six thousand years ago to being a significant dimension of nearly all complex organizations. Much of the evolution of bureaucracies has been driven by a concern on the part of bureaucrats for greater efficiency, for example, the development of new surfaces on which to record information by the ancient Egyptians and the computerization of modern bureaucracies. Papyrus allowed the Egyptian bureaucrats to be more efficient than their Sumerian counterparts, who recorded on clay. Computers make modern bureaucrats more efficient than those of the royal French state, who recorded information with quills on paper.

All refinements of bureaucratic procedures have been justified on the grounds that the new techniques promote communal interests by enhancing efficient management of information. Many of these refinements have also made the information less accessible to other members of communities and increased the ability of bureaucrats to maintain monopolies of knowledge.

Despite having been encompassed by many different types of organizations, enmeshed in many different cultures, and having tremendous variation in the first-order technologies used by bureaucrats to record information, bureaucracies have evidenced common consequences. Some consequences of bureaucracies have enhanced human welfare, but others have not.

The centrality of the bureaucratic structures varies from regime to regime, but comparatively speaking bureaucratic constancy has been far greater than dynastic constancy. In modern democratic nations, the political party in control of the state apparatus varies from election to election and to an extent the degree of influence of each state bureaucracy varies when the party in power changes. If the Democrats have control of the state, appropriations in some areas of bureaucratic structures expand; if the Republicans have control other bureaucratic structures expand. But, comparatively speaking, bureaucratic power waxes and wanes but little with dynastic changes. In totalitarian states, political upheaval hardly creates a ripple within bureaucratic structures.

Bureaucrats have a special shared interest that reaches from the lowest to the highest ranks: maintaining the bureaucracy. Their shared interest stems not from their relationship to the means of material production nor compassion for others, but from their relationship with the larger encompassing structure. If authorities of the encompassing social structure abolish the bureaucracy most bureaucrats have few alternatives. Their skills usually have value only within a bureaucratic structure. The extreme dependent position of bureaucrats entices most to support the status quo.

In representative democracies, legislatures and bureaucracies vie for control of state structures. Legislature-centered structures are formed on the belief that greater wisdom resides with the rank and file citizens than with the state elite; bureaucracies, in contrast, rest on the belief that an elite has greater wisdom than the rank and file.

In legislature-centered states, representatives are highly accountable to their constituents and constituents have access to information about state affairs. To retain their position, representatives must manage the calls for accountability from their constituents. In contrast, the greater access to information by the bureaucratic elite frees them, to an extent, from accountability to their clients. Consequently, when legislators and bureaucrats conflict with one another, the bureaucrats are often in a more powerful position than the legislators. The position of legislatures in their confrontations with bureaucracies would be strengthened if legislatures were populated with representatives who could not be reelected and if the tenure of bureaucrats were limited.

Interaction contextualized by markets generates an instrumental mentality similar to that generated in bureaucracies, but marketplace transactions often are not monolithic. Each seller in a market where others are selling the same product is aware that each buyer has alternatives. If a merchant is to succeed, he must solicit approval from buyers in competition with other sellers or displace other sellers. Success in the marketplace requires flexibility, and each trader must have autonomy to make adjustments to changing conditions. Success in bureaucracy is contingent on eliciting approval from one's superordinates. The trader wheels and deals with a series of others and seeks out alternatives; the bureaucrat follows guidelines, attempts to impose guidelines on others, and seeks approval from his superordinate. Traders regard bureaucrats as the source of evil constraints; bureaucrats regard traders as a source of disorder.

In capitalist nations with a legislature-centered state structure, bureaucracies are commonly established to protect citizens from exploitation by commercial interest. Commercial entities are called upon to abide by the formats offered by state bureaucrats. Conflict, bribery, and collusion between the elite of the market and state elite is a common consequence.

Despite the fact that bureaucracies are always an adjunct of a larger social unit, like market-, temple-, and palace-centered structures, they merit recognition as a distinctive social form. Like temples and palaces, the network of relationships that composes bureaucracies has a definite ranking, but in contrast to temple- and palace-centered networks, the ultimate focal person of each bureaucracy is accountable to an external unit.

Many critics of modern society recognize that bureaucratization is a major social problem insofar as it undermines solidarity and compassion. Yet the growth of bureaucratic structures continues unabated. Social scientists seem to have contributed to that growth by dedicating their efforts to developing more effective bureaucratic procedures. They might concern themselves instead with formulating alternatives to bureaucratic structures.

7

Written Languages

In the past, some scholars have dichotomized human societies into the civilized and noncivilized and have equated the civilized with the presence of written languages. Recent studies of the emergence of civilizations demonstrate that noncivilized and civilized societies cannot be cleanly dichotomized (Couch 1984) and that some nonliterate societies, such as the Chimu and Inca, merit the designation of civilization. In addition, studies of the emergence of the Sumerian and Egyptian civilizations (Larsen 1988; Baines 1988) indicate that those civilizations were in place before a written language was developed. All modern civilizations, however, would collapse if their citizens lost their ability to write and read. Students, from kindergarten through graduate school, spend a substantial amount of their time mastering reading and writing and those who fail to become literate usually live on the fringes of contemporary society.

Written languages preserve information in artificial forms just as constructed configurations, calendric markings, and numeric tallies do. But whereas they all preserve a particular and limited type of information, written languages can preserve nearly all discursive information in artifactual form. Written languages were originally developed to bridge time, and only centuries later were used to bridge space.

For over four millennia messages, notes, and narratives were manually inscribed on a variety of surfaces with a brush, stylus, or quill. In comparison to printing, the manual inscription of discourse is an exceedingly laborious task, yet each written language provides a procedure for preserving information that is less time consuming and far more reliable than orality, and readers can acquire discursive information more rapidly than listeners. Printing presses, typewriters, and personal computers have partially replaced handwriting, but handwriting continues to be a significant procedure for managing information. Paper and pencil are far more portable than printing presses and marginally more portable than the latest "notebook" computer.

EARLY WRITING

At one time writing was thought to have been invented ex nihilo about 3000 B.C. by the Sumerians. The first written language probably was invented by the Sumerians, but in the preceding centuries complex tokens with lines and notches traced on them with a stylus had come into common use, temple-centered trade networks had dramatically expanded (Wright and Johnson 1975), and there had been a significant increase in the number of record keeping specialists (Nissen 1988:80). These developments set the stage for the invention of a written language in Mesopotamia.

By 3000 B.C. the city of Uruk had become the hub of a trade network with a population of at least ten thousand (Larsen 1988:179). Approximately 85 percent of the earliest texts found at Uruk reference economic goods. Most of the other tablets are inscribed with lists of officials. The earliest texts contain numeric concepts as well as pictographs that depict types of goods.

Early Sumerian writing was a supplementary procedure used in conjunction with numeric tokens by record keepers to record economic transactions. In one sense the first step toward a written language was taken when an incision on a ball of clay was substituted for a token that designated a type of good. The use of writing and tokens to record economic transactions overlapped for at least half a century (Schmandt-Besserat 1992:70). The first written sentences described movement of grains stored at temples: they included the verbs "take in" and "give out" (Biggs 1974:43). The earliest texts "describe almost exclusively economic processes" (Nissen 1988:89).

Once a technique for referencing economic transactions had been developed, the invention of signs that referenced other activities quickly followed. Most of the early pictographs for verbs were composite signs that contained two signs, each of which referenced an object. For example, the action "to eat" was written by inscribing a bowl used to portion out barley allotments next to a human mouth (Nissen 1988:85). Characters for a host of objects, activities, and qualities were created. Early elaborations included the invention of characters for items that were not traded, such as chariot, sledge, and wild boar (Schmandt-Besserat 1986:38). The larger temples established tablet houses where students were taught how to write (Logan 1986:67).

For the first few centuries writing was used only in conjunction with temple-centered economic activity. That writing could be used for other purposes was not immediately apparent. It was not until about three hundred years after the Sumerians had begun to write sentences that they rendered some their sagas into writing. The sagas "were composed when writing was either altogether unknown, or, if known, of little concern to the illiterate minstrel" (Kramer 1959:202).

Writing diffused rather rapidly into other cities in the region. Literate record keepers appeared in Elam shortly after the Sumerians had invented writing (Hinz 1973:28), and a written language was developed by the Egyptians within two or three centuries. The most ancient written documents were cryptic notations. For example, one document from ancient Elam contains drawings of horses' heads of three different types each in conjunction with a numeric notation. A reasonable interpretation of this document is that the pictographs of horses with bristling manes depict stallions, those with drooping manes depict mares, and those without manes depict foals. Pictographic characters next to each set of drawings of horses probably reference the owners of the horses (pp. 29–30).

Writing seems to have emerged among the Egyptians in conjunction with palace-centered activity. "Almost all the contexts from which early writing [in Egypt] has been found are royal" (Baines 1988:196). The Egyptians did write continuous texts until about two hundred years after they had developed a written language. The first written narrative does not appear in Egypt until about 2600 b.c. and "it is very simply phrased" (Fisher 1989:70).

When the Chinese and Maya developed written languages, for the first few centuries they also used them only in conjunction with only a few activities. In China early writing was used to record eclipses, dates, predictions of military campaigns, and sacrifices to ancestors. The Maya used writing to record celestial events and political achievements.

Early literacy was limited to a small percentage of the citizens of ancient literate civilizations. Perhaps no more than a few dozen residents of each Sumerian city could write and read. Baines (1982) estimates that in Egypt during the Fourth Dynasty not more than 1 percent of the Egyptians were literate. Nonetheless, writing had consequences for the nonliterate as well as for the literate.

PICTOGRAPHIC WRITING

Many different types of written languages have been developed, but the two basic types are pictographic and phonetic. No mature written language is solely pictographic, but the most ancient written languages were based on pictographs, i.e., depictions of objects. These inscriptions indirectly referenced discourse, and as pictographic writing matured, characters that referenced sounds were developed and written languages became more isomorphic with speech.

Even with an elaborate set of pictographic symbols it is difficult to render discourse into written statements that can be transformed back into discourse with fidelity by a second person. A Sumerian proverb noted "a scribe whose hand matches the mouth, he is indeed a scribe" (Green 1981:359). A number of techniques were developed to make pictographic languages more isomorphic with spoken languages. One technique was composite characters. For example, the Chinese pictograph 0 designated the Sun and the pictograph o designated the Moon. "Subsequently the ideograph '0o' was used to designate shining" (Couch and Chen 1988:158).

The principle of homophony also was employed by the Chinese, Egyptians, and Maya. All spoken languages have words that sound the same but have different referents. The English words *see* and *sea* are homophones, as are the words *eye* and *I*. The literates of the early writing societies exploited homophony to make their written languages more isomorphic with speech. If this system were used by English speakers, the pronoun *I* would be written by drawing a picture of an eye; the verb *see* would be written by using the character that referenced *sea*. Whether a depiction of an eye referenced an eye or the pronoun "I" had to be inferred from the context.

The earliest pictographic characters were composed of easily recognized configurations. But as pictographic languages matured, the literate specialists developed stylized characters that could be recognized only by those who received special training and the characters became more evocative. Calligraphy flourished, in the course of which writing and art were almost completely amalgamated. In ancient Egypt the same word was used for writing and painting; among the Maya the sign for "to write" also meant "to paint". The elaborations made it possible to more adequately transform speech into writing, but they also made written information less accessible. The distinction between the literate and nonliterate was a major chasm in all ancient pictographic societies.

Refinements of written languages followed a different trajectory when writing matured in a market context than when it matured in a palace context. In Mesopotamia, where writing evolved in markets, there was movement "toward making writing easier and more universally useful" (Nissen 1988:136). Characters were created for syllables, and the development of a syllabic language in Mesopotamia was probably stimulated by "pressure to do more than reproduce only Sumerian [speech]" (p. 139).

In contrast, writing was used in Egypt and China from the beginning to administer affairs of state. In both instances, a large number of complex characters were developed and writing became a tool of the state elite. Two simpler systems of writing were subsequently developed in Egypt, but throughout the three thousand years of Egypt's existence, its official written language remained the elaborate pictographic system developed in the early dynasties.

The Egyptians, Chinese, and Maya endowed their written languages with sacredness. The literates of these societies also conceptualized writing as an instrument to be used to maintain social order. The Egyptian god Thoth was believed to be both the inventor of writing and charged with the obligation of maintaining social order (Senner 1989:11). In China, written statements that were to be discarded were placed in baskets on walls and burned on the birthday of the inventor of writing until well into the twentieth century.

Oral specialists probably had little interest in written languages when they emerged. Ancient oral specialists probably were even less impressed with pictographic writing, and from their point of view it was not worth the effort to master it. When societies became transformed from nonliterate to literate ones, though, it seems that orality and writing coexisted for at least a few centuries in most cases. In at least some instances during the early stages of literacy, orality was used to preserve sagas and writing to preserve administrative information.

When writing became pervasive, some of the sagas were transformed into written form. State elites probably initiated the translation of oral sagas into written accounts. In Sumer, sagas became written as palace-centered states were established. Two millennia later, when phonetic writing entered Greek communities, a tyrant of Athens initiated rendering the *Iliad* and *Odyssey* into writing (Jensen 1980:97). When information that had previously been preserved orally began to be preserved in writing, oral technologists lost their elite standing.

Despite the unwieldiness of pictographic writing, it provides a more effective procedure than orality for preserving information. Pictographic specialists can inscribe information without memorizing it and the inscribed information endures longer and with greater constancy than orally preserved information.

As pictographic writing replaced orality as the dominant procedure for preserving information, the content of the information preserved underwent a number of subtle changes. It came to reflect sentiments of the temple and palace elite more than did orally preserved information. Temple and palace celebrations gradually changed from gatherings informed by orally preserved information wherein all were highly responsive to dramatic performances into rituals where a literate elite offered enduring truth to the nonliterates. As rituals informed by written statements replaced celebrations informed by oral compositions, the communal gatherings became more orderly and less evocative. Apollo replaced Dionysus.

The transformation of exciting joyful celebrations into ritualistic readings of a sacred text was incremental and subtle, and seems to have occurred in conjunction with rendering narratives into written form. For example, in late imperial China on the first and fifteenth day of every moon, officials met in shrines and at the command of the master of ceremonies an orator read from a sacred text. Neither the officers nor their attendants paid much attention to the reading. They amused themselves "with tea and tobacco in one of the side apartments" and the audience consisted "almost wholly of vagrants, idle people who were loitering about the place, beggars, and truant boys" (Mair 1985:353).

The maturation of writing, at least in the Old World, also was associated with a transformation in the gender of divine figures. "A culture of major female deities accompanied by minor male deities gave way to one of major male deities accompanied by minor female deities" (Lind 1989:89). The communal ceremonies were modified to promote the interests of the warrior elite. "There was intense activity around mythic revision of attributes of powers of deities" (p. 90) as gods replaced goddesses. In Uruk, the dominant city of Mesopotamia, the city goddess, Lady of Eanna, was supplanted by the god An, who was situated in a gigantic new temple complex as writing became the dominant technology (Nissen 1988:100). Simultaneously, segments of oral compositions that elevated females were selectively deleted and the exploits of male warriors were elaborated. A similar transformation probably occurred in the early Egyptian dynasties. The most ancient depictions of Ma'at, the Egyptian divine being of truth and order, are as an adolescent girl (Rice 1990:83).

The emergence of coalitions of literates and warriors that culminated in patriarchal state structures involved more than changes in the dominant procedure for managing information: "On the battlefield where the male defenders clashed, men of the matrilineal order defended their customs as fiercely as did men of the patrilineal order" (Lind 1989:96), but in the long run the proponents of the male deities prevailed.

The deliberate manipulation of information was facilitated by writing, but the information specialists of some nonliterate societies also manipulated information for propagandistic purposes. For example, the Inca maintained two accounts of how the state elite come to power. One, which circulated only among the state elite, offered an account of how the Inca came to power via conquest. The account recited at state sponsored festivals claimed that the Inca elite had descended from Gods (Gibson 1948:47). Written narratives are more amenable to reflective analyses than oral ones, but the deliberate manipulation of information certainly is not limited to written texts.

In general, the literate elite of pictographic societies controlled the flow of information critical for administrating communal affairs more completely than the record keepers of oral or phonetic societies. During times of stability, the bureaucratic elite of pictographic societies more than the monarchs programmed state affairs. During times of instability, especially when a nation was attacked by a foreign military force, usually the interest of the warrior elite come to the fore and the influence of the literate elite faded. More often than not, within a generation or two of peace the literate elite, especially the bureaucrats, reasserted their dominance. Except for relatively short periods of turmoil in those ancient civilizations where pictographic writing was the dominant information technology, the literate elite exercised hegemony.

PHONETIC WRITING

Phonetic writing was invented only once, and it was a cross-cultural process. A major step toward phonetic writing occurred when the nonliterate Phoenicians learned of writing from the Egyptians and began using some of the Egyptian pictographs that referenced sounds to reference some of the phonemes of their spoken language. However, the Phoenicians did not invent inscriptions for their vowels. Consequently, some claim that the first truly phonetic system was not developed until the Greeks became literate about 700 B.C. (Carpenter 1933, 1938; Havelock 1982). The Greeks' invention of signs for vowels allowed for the reproduction of spoken discourse with far greater fidelity than any previous written language.

It is only necessary to learn twenty-five to thirty-five phonetic characters to transform spoken discourse into written discourse. Once the character for each sound has been learned, then it is hypothetically possible for a person to render all discourse into inscriptions with a fair degree of fidelity. All phonetic writing prior to that of the Greeks, as with pictographic writing, was capable of reproducing only "muffled discourse" (Havelock 1982).

Just as the Phoenicians had earlier acquired awareness of literacy from the Egyptians through trade contacts, the Greeks learned of Phoenician writing through trade contacts. The Phoenicians had two sets of phonetic characters. One set was used for commercial activities; the other for monumental inscriptions. It was the commercial characters that entered the Greek communities. Neither a temple nor a palace elite controlled writing in most Greek communities, and it was not until after literacy became pervasive among the adult males that the Greeks began using writing for bureaucratic purposes. For example, the Athenians did not use phonetic writing to record communal resources until 456 B.C., and then only to a limited extent (Hammond 1967). The Greeks originally used phonetic writing to inscribe rhythmic couplets (Havelock 1982). All of the earliest texts of the Greeks are poetic.

Book-length Greek prosaic compositions did not appear in Greek communities until about 400 B.C. (Havelock 1971). That was about three hundred years after phonetic writing had become available for the preservation of discursive information. In the intervening period, oral technologies and phonetic literacy coexisted.

The establishment of phonetic writing as the dominant procedure for preserving information is credited by some for stimulating the emergence of democracy among the Greeks. But democratic state structures only emerged in Greek communities where phonetic writing entered via markets. In Crete, where phonetic writing was introduced by a palace elite, a democratic state structure did not emerge. In Crete "[t]he purpose of law codes was not to make public, and thereby 'democratize' the laws, but to mystify them" (Stoddart and Whitley 1988:766). In Athens "public inscriptions make a late appearance, and if anything were a consequence rather than a cause of democracy" (ibid.). Perhaps no more than three hundred people were literate in Crete whereas in Athens, literates "must have numbered a few thousands" (ibid.). Between 700 and 450 B.C. only five graffiti writings have been found in Crete, compared to over eight hundred from Athens (pp. 763–64).

Literacy may have contributed to the establishment of democratic state structures in some Greek cities, but if such was the case it was unrecognized by the Greeks themselves. Despite the extensive written evidence from ancient Athens, there is no evidence that literacy was ever "a popular cause or a subject which interested democratic politicians" (Harris 1989:62).

The maturation of phonetic writing was marked with greater fluidity than the maturation of pictographic writing. For example, during the period when writing was supplanting orality among the Greeks, some writers made a letter one way and others made it a different way; some used one letter for a particular sound; others used a different letter. How words were spelled varied from individual to individual. Neither spellings, syntax, nor grammar were standardized in Athens until about 400 B.C. The idea that phonetic writing should have a definite format was not widespread in Europe until about 1500 A.D. The first grammar for a vernacular language of western Europe was written in 1492 (Karttunen 1982:396) about forty years after print entered Europe. In contrast, the early literate elite of Egypt and China developed elaborate formats for their pictographic languages.

The pictographic languages of Egypt and China matured in state bureaucracies and for a few centuries they were most commonly used to keep records. In contrast, phonetic writing matured in Phoenician and

Greek markets, and among the Greeks it was used from the beginning to render poetic narratives into written form. Writing was frequently used to enhance oral performances in many Greek communities. For example, Herodotus first wrote his histories to deliver them as speeches at the Olympic Games. Only later were narratives written for readers.

Later the state officials of the Roman Empire and the priests of the Catholic church employed phonetic writing to administer affairs of the state and temple. When a society with a written language conquered a nonliterate one, the conquerors often used writing to administer the conquered and to co-opt leaders of the conquered into the state structure of the conquerors. Writing entered the Arabic world in the sixth century of the Christian era and the Koran was composed the following century. Shortly thereafter the sword and pen were combined to bless millions of nonbelievers with Islamic dogma. Christianity spread in much the same way.

Modern literates presume that the terms *speaking* and *writing* reference two different forms of communicative activity, as do listening and reading. Those dichotomies were not clearly made in most preprint societies, where writing was largely a supplement to speech and was not regarded as a form of communication that had an existence independent of vocal communication. In societies with low rates of literacy, most people obtain information preserved in writing by listening, not by reading.

In situations as diverse as among the bureaucrats of ancient China, the sophists of classical Greece, and the officials of the Catholic church of the Middle Ages, most writers / readers also spoke to one another and the distinction between listening and reading was not clearly conceptualized. The dichotomy of listening versus reading emerged among the Greeks about 400 B.C. when many began to write prose instead of poetry. Among the literate elite of Rome during the time of Julius Caesar and Augustus, listening to a reader instead of reading for oneself was a common activity. When Augustus could not sleep he summoned a reader "instead of reaching for a novel as a modern person might" (Harris 1989:226).

Until after the invention of printing, nearly all Europeans acquired awareness of the content of the Bible by listening to a literate priest. The dissemination of information via speech supplemented by writing continues in all modern societies. Modern state authorities, priests, political candidates, and literate elite compose speeches in writing to be delivered orally. Such performances are essentially speeches supplemented by writing. It is only in postprint societies with high rates of literacy that a significant percentage of the population has acquired large quantities of information by reading and conceptualized listening and reading as two distinctly different types of activities.

Phonetic languages are far easier to master than pictographic ones.

Several reformers have called for replacing the Chinese pictographic characters with a set of phonetic characters. Chinese students of science complain they spend more time learning to decipher and inscribe the pictographic characters than they spend mastering scientific concepts. Others defend the continual use of pictographic characters for their esthetic merit and resist the vulgarization of Chinese writing.

Arabic script is somewhat more cumbersome than Roman script. One estimate is that when Turkey replaced Arabic script with Roman script, children mastered the new script in half the time (DeFrancis 1989:173). Japanese writing is a syllabic system derived from Chinese writing (Gaur 1984:34). One estimate is that it takes Japanese schoolchildren two years longer to learn to write and read than schoolchildren who learn a phonetic system. The high rates of literacy achieved in Taiwan and mainland China in the late twentieth century, however, demonstrate that high literacy rates can be achieved with a pictographic language despite the extra effort required.

A number of phonetic written languages were developed during the nineteenth century as Western civilization diffused into nonliterate societies. In nearly all cases these systems evolved from a pictographic script with elements of phonetics into a syllabic script, only to finally be replaced by a phonetic script (p. 131). Once a phonetic script has been developed, adults are usually capable of mastering it within a few weeks or at most a few months. For example, among the Vai, who developed a phonetic script, "on the average two to three months of lessons were required to achieve functional literacy, although some persons admitted to needing more time" (Scribner and Cole 1981:66).

Pictographic literacy fostered continuity across time, dichotomized the citizenry into the literate versus illiterate, enhanced monopolies of knowledge, and invited the rise of totalitarian governments. Phonetic writing renders speech with greater fidelity, promotes the sharing of information, makes it more difficult to maintain monopolies of knowledge, and lessens the chasm between literates and nonliterates. Monopolies of knowledge in phonetic societies are more likely to suffer leakages than those of pictographic societies. But phonetic written languages have only facilitated democracy when a high percentage of the population has been literate and state authorities did not control the dissemination of information via written languages.

In oral civilizations, information specialists were in a more powerful position than in literate civilizations. The ruling elite of oral civilizations were dependent on a very few oral specialists. In literate societies there usually are a large number of literates and if one scribe does not perform as a state authority wishes him to he can easily be replaced. It is difficult for an individual to establish himself as an indispensable source of information for state authorities in literate societies, especially in phonetic societies.

The monopolies of knowledge maintained in nonliterate societies usually were not monolithic, because they usually were topic specific. The Inca were an exception to this generalization: they constructed multifaceted monopolies of knowledge despite being nonliterate. More commonly, monopolies of knowledge in nonliterate societies were topic specific. For example, among the South Seas Islanders, some information specialists mastered navigational poems, others memorized poems containing calendric information, and still others memorized genealogies. In contrast, the monopolies of knowledge maintained in pictographic societies often were multifaceted.

Many of the transformations of social structures that accompanied the maturation of writing as the dominant procedure for preserving information were so gradual that many who experienced the changes were unaware of them. For example, literacy is associated with profoundly different types of judicial procedures than is orality. The judicial processes in oral societies are more public, fluid, emotional, and varied than they are in literate societies. Prior to the replacement of orality with phonetic writing as the dominant information technology in ancient Athens, trials were communal extravagances wherein prosecutors and defenders competed for approval as an audience expressed its appreciation of their efforts. In contrast, among the ancient Egyptians of the Middle Kingdom the plaintiff and defendant offered written statements to a judge so that the judge would not be influenced by rhetorical or personal matters. Access to the judicial process for nonliterate Egyptians was only available if they could employ a scribe. Egyptian law reflected the referential bias of writing, while Greek law of the preliterate period reflected the evocative bias of orality.

QUALITIES OF WRITTEN INFORMATION

Written languages vastly increase the amount of information that can be preserved. A literate specialist has access to hundreds, if not thousands, of books whereas an oral specialist had, at best, access to tens of poems. The amounts of information modern literates have access to are so much greater than that available to oral specialists that the differences are comparable to the amount of wheat that can be harvested by those using a combine in comparison to those using a hand sickle. In addition, written languages allow for the more efficient processing of information. Proficient readers read faster than the flow of speech. Furthermore, readers can skim while listeners cannot.

Despite the greater power of written languages to preserve information, groups of oral specialists as diverse as the Druids of Briton, the navigators of the South Seas (Makemson 1941), and Brahmin priests (Gaur 1984) resisted writing. Some oral specialists regarded writing as a threat to their elite position and resisted writing for that reason. Others imbued in an oral tradition, such as Socrates, thought writing preserved only a shallow knowledge, that it would destroy the ability of humans to remember, and that thought written languages would erode compassion for one's fellow human beings.

Advocates, however, have claimed that written language opens new vistas for humanity. Sequoia, who developed the Cherokee script, noted, "much that red men know they forget, they have no way to preserve it. White men make what they know fast on paper like catching a wild animal and taming it" (as quoted in Harbsmeier 1988:265). The spoken word, even in poetic form, is fleeting; what is written endures.

When ordinary speech and poetry were the dominant means of preserving and transmitting information, each community member was exposed to much of the same information, but as written languages replaced poetry more information became secretive and some community members avoided exposure to some information. When information is dispensed at public ceremonies by oral specialists, it is difficult to avoid exposure to it. When information is dispensed in books, each person must seek information to obtain it.

Most literates regard the written word as having greater substance than the spoken word, although the contrary assessment is often offered by nonliterates. When writing came into use in western Europe, greater significance was assigned to oral statements by witnesses, testimony, than to documents. In medieval Europe "a land transaction required witnesses who heard the donor utter the words of the grant and saw him make the transfer with a symbolic object, such as a piece of turf from the land" (Zuboff 1988:78). Disputes of ownership were settled by calling witnesses to testify as to what had occurred. Written documentation was not accepted as evidence in legal disputes among the Greeks until about four centuries after literacy had entered the Greek world (Harris 1989:72–73).

In oral societies, commitments are constructed in evocative transactions; in literate societies they are constructed by signing contracts. From the standpoint of the nonliterate unfamiliar with written languages, "The spoken word was connected to the inconvertible realities of body experience, while the written word was a thin, substanceless scratching" (Zuboff 1988:77). Those embedded in a literate world regard the spoken word as an undependable fleeting phenomenon. When literates desire to construct a firm commitment they put it in writing. Nonliterates shake hands.

Writing, in particular phonetic writing, depresses evocative communication and favors communication via referential concepts. Even ordinary speech is more likely to stir emotions than is phonetic writing, and listening to an oral performance is far more likely to stir emotions than reading a book. Tempo, intonation, and stress accompany all referential information preserved by oral technologies and give orally preserved information emotional substance. Pictographic writing gives greater emphasis to evocative symbols than phonetic writing, but far less than orality. Speech and orality leave sentiment; phonetic writing leaves an informing artifact. Calligraphy (a composite term one derived from the Greek words *beautiful* and *writing*) has flourished in pictographic societies.

All techniques that preserve information in artifactual form disembody information, but whereas the more ancient inscriptions preserved only small quantities of information, written languages provide a means for preserving large pools of disembodied information. Readers can acquire large pools of information from books; listeners can acquire large pools of information only from people. Authors can be unavailable to readers; speakers, in contrast, are always available to listeners. One can throw a tomato at an offensive speaker; usually it is difficult to throw a tomato at an offensive author. Information obtained from a speaker is always linked to a person. Even the tyrant can be shouted at when he speaks to his victims. Written languages provide a procedure whereby one can communicate without becoming accountable. Examples of such communication are graffiti, secret ballots, and anonymous letters.

When information is shared via writing, the writer and reader seldom share an immediate context, whereas speakers and listeners nearly always do. When communication occurs via speech, the participants are mutually aware that they share an environment. When separated people communicate via writing, they may have a common consciousness derived from the content of the written information but they do not have a mutual consciousness based on an encompassing environment.

Orality and speech provide dynamic ephemeral auditory information; writing provides static enduring visual information. Spoken information only exists in process, as movement; written information exists as a series of objects and is static. Even chalk notations on a blackboard are more enduring than speech.

Written information is more likely to be subjected to reflection than spoken information. Consequently writers and readers are more likely than composers of and listeners to oral narratives to note internal inconsistencies. Extended oral compositions, such as the *Iliad*, are more likely to contain inconsistent statements than are extended written compositions.

The production and reading of written statements invites an analytic standpoint, whereas the production of and listening to oral compositions invites expressive responses. Readers are more likely to focus on content than are listeners; listeners are more likely to emote with and focus on performers.

Orally preserved information does not remain constant when it is the object of reflection. If two persons are to analyze orally retained information, they must reproduce the information in speech and the information is likely to change as they reproduce it. In contrast, a plurality of literates can make written information the object of reflection and the information remains constant as they examine it. Rationality emerged and replaced emotional insight as the source of knowledge as phonetic writing replaced orality as the dominant technique for preserving information.

Speakers note the responses of listeners as they offer information; in contrast, writers assess the imagined responses of readers. Speakers can assess the level of understanding of listeners in the immediate situation; writers cannot. Teachers more or less continually assess the understanding of students when knowledge is transmitted via speech. On the basis of their assessment of the level of understanding of the student they proceed on or loop back and repeat themselves or offer the ideas in different words. Speakers can take remedial measures to facilitate shared understanding in the middle of a communicative act; writers cannot. One consequence is that writers usually are more concerned with format than speakers.

The interaction between oral specialists and their audiences was fluid, dynamic, passionate, and simultaneously reciprocal. The interaction of writers and readers is accomplished via enduring artifacts by staid persons acting in solitude, and reciprocity, if present, is delayed. If passions are stirred in writers or readers they typically are solitary passions. The passions stirred at oral performances are shared.

Written languages, especially phonetic ones, facilitate the formation of abstract concepts and propositions. The movement to greater abstraction occurred among the Greek intellectuals as they became literate. Looking at something became the ultimate intellectual operation as the Greeks were changed from an oral to a literate society, and logic emerged as a distinctive intellectual activity as they began to compose prosaic narratives. The miracle of Greece occurred between the time literacy entered Greek communities and when prosaic literacy became common.

Knowledge became more rigid as literacy replaced orality. The fluidity of orality made it difficult to establish a fixed orthodoxy. The literate elite have often established a particular version as the valid one and those who do not espouse the orthodoxy become heretics. Written languages facilitate the construction of "awe" "full" knowledge.

Written languages have also allowed for the emergence of the belief that knowledge is something that exists independent of human action and thought. One example of that conception of knowledge is Plato's claim that true knowledge only exists in the realm of ideas. Plato conceptualized knowledge as everlasting, constant, and independent of sensate experiences. Many modern logicians adopt the same standpoint.

MYTHS AND HISTORIES

Most nomadic groups maintained only limited accounts of the past that had little chronological depth. One group of Australian Aborigines, for example, divided the past into three stages: (1) events witnessed by the speaker, (2) a long time ago, and (3) the creation period. A few nonliterate societies constructed oral genealogies that reached twenty to thirty generations into the past, but these chronologies were fluid and durations, if noted, were vaguely specified. Several nonliterate civilizations constructed elaborate accounts of the past. In a few instances, such as the preliterate Greeks and the Incas, some oral specialists known as remembers were charged with preserving narratives about the past. A few nonliterate civilizations chronologized the past, but most oral accounts of the past had little chronological depth and durations usually were not precisely specified. History began when people merged written narratives with calendars. That merger allowed for the construction of extended chronologies that contained precise measures of durations.

The construction of extended chronologies and precise durations usually did not flower until a few generations after a society became literate, although on several occasions after a nonliterate society established contact with a literate one some members of the nonliterate society attempted to construct a history of their society shortly afterwards. For example, after a Native American learned of writing, he "gathered the names of many years from the old people and placed them in chronological order as far back as he was able to learn them" (Mallory as quoted in DeFrancis 1989:36). Other members of his society were no more appreciative of his effort than modern high school students appreciate the efforts of modern historians.

Shortly after literacy emerged in the Mesopotamian and Nile valleys some literates began to record the celebration of holidays and the accomplishments of the elite, but it is questionable that these accounts merit the designation of history. The narratives evidence "little concern with specific dates or proper chronological order" (Van Seters 1983:299). The most ancient histories of the Egyptians sequenced the past by year-names, not numerically. The year-names were replaced with a numeric chronology during the Second Dynasty. These earliest Egyptian chronologies were constructed for administrative purposes, not to replace oral compositions (Baines 1989:133). Oral narratives about the past continued to be composed by Egyptians for several more generations (Foster 1980).

Subsequently the Egyptian literate elite merged palace day books, written accounts of military affairs, and astronomical observations with orally preserved narratives to create an official history (Van Seters 1983:293). The earliest historians did not offer explicit assessments of past events but they nonetheless implicitly rationalized the extant social structure. There is no evidence that Egyptian histories were deliberately constructed to rationalize the rule of a particular monarch or lineage until the Twelfth Dynasty, when a usurper became king. That usurper set out consciously to win the support of the whole land with a "diverse program of literary propaganda" (p. 173). Later Hatshephut and Harembab, both of whom instituted new dynasties, also had histories written that attempted to legitimate their rule.

As writing became the dominant information technology and palace-centered states were put in place, literate tutors attached to temples and palaces constructed histories that rationalized the special position of the state elite (Lind 1989:91). These ancient histories equate the beginning of order with the founding of their nation. According to ancient Egyptian histories, prior to the founding of Egypt all was chaos. Most of the non-literates were indifferent to these compositions, but nonetheless those histories contributed to the establishment of exploitative social orders.

Pictographic specialists frequently constructed histories that ordered the past in cycles. Cyclical histories indicate that every present and future is a recurrence of past events. That frame of thought seems to have been an extension of the frame of thought originally formulated by time-keepers to characterize cycles of celestial phenomena. These ancient historians apparently deduced that if celestial events were cyclical, so were human affairs.

Those who composed cyclical histories must have been aware that the current social structures were not merely replicas of past ones. However, the ordering of the past in cycles probably was not the consequence of deliberate attempts to compose fraudulent accounts of the past. It is likely that the Egyptian men of knowledge firmly believed that it was "unthinkable that nature and society should follow different courses, for both alike were ruled by Ma'at—(right, truth, justice, cosmic order)" (Frankfort 1948:277). Their commitment to a cyclical frame of thought may have been so complete that they could not conceive of the possibility of emergent phenomena. The Maya information specialists, like the ancient Egyptian intellectuals, also concluded time was without change and sacred (Stross 1989).

Some have characterized civilizations that developed cyclical accounts of the past as ahistorical (Toulmin and Goodfield 1962; Fehl 1964). But those ancient chronologizers ordered the past sequentially, offered measures of duration, and described human affairs as surely as do linear historians. They merely denied transformations, and presumed that nature and society existed in a steady state.

Some ancient cyclical narratives of the past contained elements of linear history. The Chinese historians, for example, recorded inventions and other novel occurrences. Conversely, elements of cyclical histories have had currency in phonetic societies. Aristotle, for example, thought he could be living before the Trojan War as well as after it. But, in general, as people began using phonetic writing to construct accounts of the past history became linear.

Linear historians focus on transformations instead of redundancies. For example, whereas ancient Egyptian history stressed constancy, Jewish history described cataclysmic events that occurred only once, such as their escape from Egypt. The Jews also were the first to construct an extended history wherein events of significance for the people took precedence over the activities of the kings (Van Seters 1983:355).

The Greeks became concerned with chronology and duration as they became literate. Homer's poetic account of the Trojan War gives some attention to chronology and durations. Herodotus's prosaic history of the Persian Wars displays more concern with chronology and durations. Thucydides, who followed Herodotus by only a generation, carefully chronologized his history of the Peloponnesian War and offered precise measures of duration.

The beginnings of many ancient histories were temporally anchored to a cataclysmic transformation linked to a palace elite such as when a ruling lineage was installed. Other histories, especially those of phonetic societies, began with a cataclysmic event that affected a whole people. Jewish history is anchored to creation in 3701 B.C.; the Moslems reckon from the flight of Mohammed from Mecca (622 A.D.); Christian histories are anchored to the birth of Jesus. The custom of using the birth of Jesus as the temporal anchor was first proposed by a priest in 525 A.D. (Needham 1969:286). Civilizations as varied as the Hindu, Chinese, Maya, and European have extended chronologies indefinitely into the past. Some have done that with great precision, if with doubtful authenticity. According to one Hindu history, between 7584 and 7308 B.C. the temperature of the Earth rose and created heavy vegetation (Prasad 1989:150).

As writing replaced orality, third-party referential descriptions replaced evocative oral performances as the primary source of information about the past; precision supplanted passion. The performances offered by oral specialists created a shared adventure for the audiences. In contrast written histories are staid descriptions written in solitude and read by solitary readers. The past became more precise but less interesting as it moved from orality to literacy.

Almost without exception literates regard written accounts of the past as superior to those offered by poets. Thucydides, who offered the first Greek account of the past firmly lodged in the written tradition, called on historians to ignore the exaggerated fancies of poets and offer truthful accounts instead of accounts that please the ear. The dichotomy of the fictive versus the factual is recognized in all societies, but that dichotomy seems not to have been applied to narratives about the past to any great extent until after people began to write histories. On several occasions two orally composed narratives about an episode in the past that offer inconsistent descriptions had currency in a nonliterate society without generating concern.

The histories of the ancient Egyptians and Chinese were highly ethno-centric; all international contacts were assessed from a nationalistic per-spective. The adoption of a third-party standpoint that incorporated the standpoints of antagonists emerged as multiple written accounts of the past acquired currency within a society. Thucydides noted that historical accounts often reflected the interests of one side or the other and claimed that his history offered a true rendition of what had transpired. Nearly all modern historians make the same claim as Thucydides, yet nearly all histories are contaminated with ethnocentrism.

Modern historians modify accounts of the past when new information is uncovered and when a new paradigm is applied to old data. Such modifications are self-conscious acts, but the modifications of oral ac-counts of the past often were unintended. Oral performances were judged primarily on the basis of their entertainment value, and little critical atten-tion was given to the accuracy or constancy of the recitations. Oral spe-cialists who offered accounts of the past were more concerned with eliciting approval from audiences than with accuracy.

Even when audiences of oral performances concerned themselves with accuracy it did not necessarily contribute to the accuracy of the accounts. Most nonliterates presume that present conditions prevailed in the past. For example, the nonliterate farmers of a rural area in Portugal found "it difficult to believe that there was a time when Maize was not dominant in the area" (Pina-Cabral 1989:60). Consequently, when a oral performer offered a description at variance with current conditions the audience would indicate that they doubted the veracity of the description. One unintended consequence of assessing the veracity of descriptions of the past by comparing them to current conditions was the subordination of descriptions of the past to the present.

In societies with elaborate pictographic accounts of the past, the past is commonly regarded as a preferred state of being by the literate elite and the future is conceptualized as closed. In contrast, when phonetic histories became common, the present is seen as preferable to the past and the future is conceptualized as open.

When literacy is limited, direct access to accounts of the past is restricted to a small number of people closely affiliated with temples or palaces. Such people have frequently allowed their concern with preserving their special position to impact on the content of their histories. A concern with accurate accounts of the past is more likely to arise in societies where a large number of people, including some not affiliated with the palaces, are literate.

In oral societies, narratives about the past were known by nearly all adults. In literate societies only literate people had access to the written narratives. Even in societies with universal literacy, the chasm between those who write histories and those who read them is greater than the chasm between reciters of sagas about the past and audiences in nonliterate societies.

Many citizens are indifferent to history. Nonetheless, history informs their actions as well as the actions of professional historians. Most of those who make the most extensive use of historical information occupy central positions in the social structures. State officials, educators, and temple elite often rationalize their actions by referencing history.

Just as consciousness of the past changed with the emergence of history, it is changing again as recording technologies are replacing writing as the dominant procedure for preserving information about the past. A concern with chronology is receding and passion is coming to the fore as film and video recordings replace books as the primary means for preserving accounts of the past.

CONCLUSIONS

The ability to preserve and disseminate information via written languages was contingent on generating knowledge of how discourse could be rendered into inscriptions. In one sense, human beings began to develop written languages when they began making depictions tens of millennia ago, but that activity did not mature into the ability to render discourse into inscriptions until about 3000 B.C. The movement from the nonliterate world to the literate world has been marked by many starts, bursts, stops, and retreats.

Written languages transformed discourse from the auditory mode to the visual mode and created a hybrid language. The hybrid nature of written language is indicated by such phrases as "I see where the newspapers say. . . . " That hybrid is the foundation for a consciousness that favors scrutiny and reflectiveness and disfavors passion and memorization.

The transformation from nonliterate social orders to literate ones was accomplished via a series of incremental steps, few of which were thought of by those involved as drastically transforming humanity. Yet the differences between literate and nonliterate worlds are as profound as the differences between the liquids and gases in the world of chemistry. The amount of information at the command of the early writers was less than that retained by the contemporary oral specialists, but within a few generations the amount of information literates had at their command far exceeded that of the most skilled oral specialists. Small-town libraries preserve far greater amounts of knowledge than what was retained by the most skilled ancient poet.

Paradoxically, written languages eroded the ability of individuals to memorize. Memory is relevant among literates, but instead of remembering content, literates need only remember how to write and read. Efforts by oral specialists that were spent on memorizing content can be spent by literate specialists accumulating information.

Prior to written languages, several information technologies existed more or less autonomously, with each technology used to preserve a particular type of information. Numeric tallies were used to preserve quantitative information, monuments were used to preserve calendric information, and orality to preserve narratives. As written languages matured, these separate streams of information were merged. That was sometimes accompanied by the creation of centralized all-encompassing monopolies of knowledge and coalitions of literate and warrior elites to institute totalitarian state structures.

Literacy often has been effectively used to serve the interests of temple and palace elite. Civilization has been imposed on nonliterate societies by literate palace-centered societies. The Chinese used a combination of literacy and arms to bless many with the benefits of their civilization. The ancient Greeks, Romans, and Western Europeans did much the same. The nonliterate recipients of the benefits of civilization have had only the choice of compliance or disaffiliation.

Whereas orality offered embodied streams of information, written languages offer disembodied pools of information. As humanity left the age of orality and entered the world of writing, information changed from fluid streams of limited size to frozen, ever-expanding pools. In many instances knowledge became sacred. Truth came to be inscribed referential concepts and abstract propositions promoted by a literate elite. Emotions became, if not antithetical to truth, at least irrelevant. As writing replaced orality life became less exciting and more orderly.

CHAPTER

8

Printing Books

In most preprint literate societies, only people affiliated with temples or palaces had access to books. Bookstores were almost unknown, and with few exceptions libraries were adjuncts of temples or palaces. In a few instances, such as the library of Alexandria, a large number of books were accumulated at a single location. Julius Caesar proposed a public library for Rome, which was built after his death. The preprint Chinese also established state-operated libraries.

In a few preprint phonetic societies, books achieved a limited circulation among people who were not affiliated with temples or palaces. Most adult male Jews were literate (Millard 1972), and by 400 B.C. a few books circulated among the literate in Greek cities. Plato noted that copies of Anaxagoras's books could be purchased in Athens (Harris 1989:85). A minor book trade developed in Rome in the first century A.D. Books by Seneca, Caesar, Virgil, and Horace were reproduced by literate slaves. Books were "sometimes produced in script 'runs' of 500 copies or more" (Graff 1987:28).

Prior to the development of the printing press in Europe, a modest book trade had emerged, especially in the large commercial cities with universities, but ownership of books was restricted to a very small minority. After European commercial printers began to distribute printed books, large numbers of books circulated independently of networks linked to temples or palaces.

The maturation of the book-printing industry in Europe was associated with several of profound changes. Although printing had appeared a few centuries earlier in China, its maturation was not associated with any major transformations in the social structure of China. The printing of books was a state monopoly in China. In Western Europe, however, printed books were produced and distributed by entrepreneurs. The development of printing in Europe, therefore, shifted the reproduction of books from the patronage of temples and palaces to the commercial realm.

The transformations that occurred in conjunction with the perfection of the printing of books in Europe did not flow from any changes in the quality of information associated with printing. Mechanical print is more standardized than handwriting, but printed books have essentially the same qualities as hand-crafted ones (Boorstin 1985:515). The transformations that accompanied the maturation of printing books in Europe flowed from the tremendous increase in the number of books in circulation and from their distribution through trade networks instead of networks centered in temples.

CHINA

Books may have been in existence during the Ch'ing Dynasty (1765–1123 B.C.) in China. "Nearly 2,500 different characters inscribed on bone and tortoise shells have been deciphered from that period including a pictograph which still stands for book" (Febvre and Martin 1976:71). However, if books were present in China at that time, they remained uncommon until after paper was invented about 100 A.D. State officials controlled the production and use of paper. Even before the invention of print, state authorities occasionally mass produced written proclamations. On one occasion three hundred thousand broadsheets were distributed that denounced a threatening warlord (Wright 1979:233). Another widely distributed preprint proclamation recounted "the emperor's tireless efforts to control the barbarians and assure peace in China" (p. 234). Most of these mass-produced proclamations were single sheets that were posted on walls.

All the libraries of preprint China were under the auspices of the state. One library contained fifteen thousand manuscripts (Febvre and Martin 1976:72). Some printed information was commercially produced in China when printing by wood blocks was developed. For example, calendars were printed by entrepreneurs early in the ninth century, but in 835 A.D. the state issued an edict prohibiting the printing and distribution of calendars by private individuals (Twitchet 1983:23). The Chinese government formalized censorship procedures in 1090 A.D. (Ingelbart 1987:8).

The reproduction of manuscripts by movable type began in the eleventh century in China. "The state very quickly appreciated that printing could be used as a means of state control" (Twitchett 1983:32) and made it a state monopoly. Although the government printed a large number of books, the number of copies per edition remained relatively small. Few editions exceeded fifteen hundred copies as late as the sixteenth century (Rawski 1985:19), when perhaps half of the adult male citizens might have been literate.

State authorities, though, were not always able to completely control the production of printed material. For example, black market establishments arose that printed model essays and sold them to candidates for civil service positions, and some candidates smuggled copies of the essays into the examination halls when taking their examinations. But black market books never achieved wide circulation in China.

There was variation from dynasty to dynasty, but during most dynasties the state authorities exercised rather close control of printing and occasionally published large editions. In the Sung dynasty, about 1100 A.D., "six different editions of the Buddhist *tripitaka* were printed and distributed throughout the country" (Tsuen-Hsuin 1985:369). The National Academy published "no fewer than three hundred works including classics, histories, local gazetteers, imperial documents, manuals of calligraphy, classified encyclopedias, as well as works on medicine, agriculture, and technology" (p. 177). Books of fiction and drama were not printed in China until after 1600 A.D. (p. 175). Almanacs that specified the lucky and unlucky days could be found in most villages. Many of them were published by private black market printers despite it being a penal offense for anyone but the government to print them.

State officials distributed proclamations more frequently after printing was perfected. One elaboration was the printing and posting of lists of citizens guilty of asocial behavior: their names were "only removed when the wayward ones had reformed" (Wright 1979:242).

Printed reproductions of pictograph manuscripts are less costly than handcrafted ones, but the difference is much less than the difference between phonetic print and phonetic writing. The cost ratio of printed versus handcrafted publications in China was about one to ten (Tsuen-Hsuin 1985:373). In Europe within a few decades after Gutenberg the cost ratio from one to two hundred to one to four hundred.

Print technology diffused from China to other Asian nations as a state activity. For example, in the fourteenth century the Korean government began to use metal letters to print books and restricted printing to the royal foundry. Printed books did not become common despite the adoption of a phonetic writing by the Korean government early in the fifteenth century (Burke 1985:112). Very little Asian printing was informed with pecuniary motives. Until the twentieth century nearly all the Asian printing establishments were state operated.

EUROPE

In about 1200 A.D., Muslim invaders brought the technique for manufacturing paper to Spain. By 1300 the price of paper was about one-sixth that of parchment (de Sola Pool 1983:12). The paper was rougher and more fragile than parchment but quickly became more readily available. Water-powered paper mills were in operation in Europe by the fourteenth century.

It takes almost as long to write on paper as it does on parchment, but paper substantially increased in the number of books in circulation and a modest reading public emerged in Europe. It is doubtful if the literacy level of even urban centers with universities exceeded 5 percent during the first half of the fifteenth century (Scribner 1981:2).

In preprint Europe most books were reproduced by scribes affiliated with the Catholic church. One preprint Italian bookseller employed fifty scribes (Steinberg 1959:25). A few monarchs and wealthy nobles subsidized the copying of books. Some of the books reproduced were histories, and a few were travelogues (Marco Polo's journal of his travels was published in the thirteenth century), and some were reproductions of the writings of the ancient Greeks and Romans. Most of the preprint books reproduced outside the *scriptoria* were commercial, legal, or theological handbooks. At least four hundred copies of one prayer book were produced in 1437 (Febvre and Martin 1976:28).

The technology of print diffused from China to Europe via international trade (Tsuen-Hsuin 1985:315–19). About 1450, Johannes Gutenberg, a merchant, perfected the mass production of metal letters and cheaply produced an almost unlimited supply of them. The first European printings were modest endeavors of only a few score copies and for a few decades printed books and script reproductions coexisted: "A large number of manuscripts made during the late fifteenth century were copied from early printed books" (Eisenstein 1979a:51).

The printing industry slowly grew for the first three or four decades, and then rapidly expanded. In 1471, only thirteen towns of central Europe had printing presses; ten years later over one hundred presses were in operation from Italy to Sweden (Febvre and Martin 1976:178–80). "By 1480 the Koberger Press in Nuremberg had over 100 employees and operated twenty-four presses" (Steinberg 1959:48). By the end of the fifteenth century, nearly every city of Central Europe had one or more commercial presses. In Venice, the major commercial center of Europe, 268 printers published two million books between 1481 and 1501 (de Sola Pool 1983:13). Printing became the big business of the age and drove scribal reproductions from the market.

Several of the earliest printing centers were university cities, where "scholars, teachers, and students working in cooperation with authors and craftsmen organized an active book trade" (Febvre and Martin 1976:19). The first printing press in the city of Paris operated under the auspices of the university and was regarded as a handy means of producing texts (p. 21). The interests of the university faculty in a supply of books and the mercenary interests of printers were compatible. Nearly all university cities had a printing press by the end of the fifteenth century.

People who could not afford preprint books began to purchase them. In 1483, a commercial printer charged three florins per quinterno to reproduce Plato's *Dialogues*. A scribe probably would have charged one florin per quinterno for duplicating the same work. The commercial printer produced 1,025 copies; the scribe would have produced only one (De la Mare, as quoted in Eisenstein 1979a:46). Prior to printing, a book had roughly the same value as two cows, which means that a professor with his annual salary could buy about two books of law or maybe ten books of medicine (Cipolla 1956). Furst, who acquired control of Gutenberg's press, sold printed Bibles for one-fifth the price of those copied by scribes (Logan 1986:181).

None of the state or church authorities made any effort in the early decades to control printers: "Numerous monasteries welcomed printers into their midst and monks even did some printing" (Febvre and Martin 1976:172). The greater readability of print made printed books more attractive than script to nearly all. For the first few decades the books reproduced by printers were much the same as those previously duplicated by scribes. In 1490 a press was established in Florence for the purpose of improving the quality of Italian education. A swift succession of classics from the Greeks and Romans was issued. Those books were not reproduced with the intention of maintaining tradition. It was simply that traditional manuscripts were available and they sold well.

As the printing industry expanded, newly authored books were solicited by the printers and publications in vernacular languages became common. "Before 1500, about three-quarters of all printed matter was in Latin" (Steinberg 1959:83). That ratio was reversed within a few decades, and by 1550, thirty vernacular translations of the Bible were in circulation (pp. 87–88).

Some of the literate realized that the press provided a means for communicating to the multitude and wrote books to achieve fame. There were no copyright laws and the idea of being paid a royalty was slow to develop. The interests of the printers, of course, were mercenary. The authors sought to write for the masses and the printers sought to expand their markets, and so there was a gradual shift in the content of the books reproduced from sacred statements to secular ones.

The printing industry was made up of a series of loosely connected entrepreneurs, printers, and writers. Print shops became gathering places for authors, scholars, printers, and traders. "The first century of printing produced a bookish culture that was not very different from that produced by scribes" (Eisenstein 1979a:26), except these people gathered at printing shops and most of them were not affiliated with temple- or state-centered social structures. These market-centered associations were more fluid and less hierarchical than those of the *scriptoria.* An interest in writing and publishing books for a mass readership became a part of the ethos of these collectivities.

More books were produced in Europe in the 150 years after Gutenberg than had been produced in the preceding 1,500 years. About 1,250,000 copies were printed in the seventeenth century and about 2,000,000 in the eighteenth century. In the nineteenth century the publication rate shot up to about 8,000,000 new titles (Bagdikian 1971:10). There probably was a corresponding increase in readership.

The relationships between authors and printers and between printers and readers of books were very different from those that had prevailed among scribes, authorities of the *scriptoria,* and readers. The scribes of the European scriptoria and the Chinese printers were subordinates in enduring asymmetric social relationships. They produced to please their authorities, not to attract a mass readership. So long as they pleased their authorities, their position was secure. Commercially produced books were distributed via fleeting symmetric encounters between sellers and buyers and the sellers were in competition with one another. The printers, their employees, and book retailers were subject to the whims of the marketplace. If they produced books that appealed to buyers, they endured; if they did not, they failed.

Information, whether oral, written, or printed, distributed through palace and temple networks is encased in a network of asymmetric accountability, and the information distributed reflects the sentiments and interests of the temple and palace authorities. The distribution of written or printed information via the market separates the flow of information from accountability. If there is no intrusion into the markets by state or temple authorities, printed information is dispensed via fleeting symmetric encounters between strangers. The flow of information is separated from identities and personal accountability. When books are distributed in markets the sellers may not know or care about the identities of the buyers. Likewise, the buyers may not know or care about the identities of the sellers. Those who purchase books at markets need not be accountable to state authorities, authors, their community, or anyone else.

The separation of the flow of information from personal accountability transforms the type of information produced and distributed. Writers and printers accountable to authorities cater to the sentiments and interests of authorities. They usually do not formulate novel definitions of reality. When books are distributed in markets, the sentiments and interests of buyers are the primary determinants of what books will be distributed. When market-distributed printed information is unconstrained by state and temple authorities, the producers of information are more likely to offer novel definitions of reality than when authorities distribute printed information.

The absence of accountability, the ease of production, and the increases in amount of information available associated with the growth of commercial printing transformed the written word from a sacred artifact into a profane commodity in Europe. Prior to the printing press, literacy was largely restricted to the clergy and books were imbued with sacredness. Commercial printing of books secularized them.

THE PRESS AND CHURCH

The success of the Christian religion stemmed in part from the effective use of writing. Christians were among the first to put their writings into codex form, which allowed scribes to write on both sides of parchment. Codices can be more effectively used as a reference than book rolls and early on Christians established *scriptoria* to reproduce their sacred texts. The Christian texts did little to promote universal literacy, but they provided sacred texts for those dedicated to spreading Christianity and stabilized Christian ideology.

The Catholic church took little note of the printing industry for the first few decades. When books that contained criticisms of Christianity began to be printed in the late fifteenth century, the pope issued a Papal bulletin that "judged the printer's art to be useful when it multiplied good books but perverse when it spread wicked doctrine" (Grendler 1977:71). When published criticisms of the church became more common, Pope Alexander VI issued a bulletin in 1501 requiring all printers to obtain the approval of the church before publishing a book. The edict failed to stem the tide of books critical of the church hierarchy. In 1509 Erasmus published *In Praise of Folly*, which, among other things, ridiculed the church hierarchy. It sold well and ushered in a new era marked by widespread vicious satire directed at church authorities (Scribner 1981:42). Many had been critical of the church hierarchy long before printing, but never before had critics been able to communicate to so many.

What had been a trickle of criticisms became a torrent when Martin Luther nailed his ninety-five theses to a church door in 1517. Within a month printers in three different cities published Luther's theses. Luther was far from the first cleric to nail theses on a church door: "It was entirely conventional for professors of theology to hold disputations over an issue such as indulgences and 'church doors were the customary place of medieval publicity'" (Eisenstein 1979a:306). When Luther was called to account by the church authorities for the widespread circulation of his ninety-five theses, he pleaded, "It is a mystery to me how my theses were spread to so many places. They were meant exclusively for our academic circle" (p. 306). Whether Luther actually intended to circulate his criticisms outside the confines of the clergy is of little significance.

The press provided the means for hundreds of thousands of people, if not a few million, to have access to information that previously was available to at most a few thousand: "Between 1517 and 1520, Luther's thirty publications probably sold well over 300,000 copies" (p. 303). Between 1518 and 1525, Luther's writings accounted for one-third of all German language books sold (Anderson 1983:163). Nearly all of those who had access to books that criticized the church prior to the sixteenth century were embedded in a church-centered network. But henceforth, books critical of the church became readily available to a substantial minority as the Protestants and Catholics competed for hegemony in defining reality. Luther commented that the printing press was "God's highest and extremist act of grace, whereby the business of the Gospel is driven forward" (as quoted in Eisenstein 1979a:304).

Vulgar criticisms were commonly published. One Protestant pamphlet was entitled *Pope Donkey and Cow Monk;* a friar offered *The Lutheran Madman* (Febvre and Martin 1976:291). One printed cartoon depicted mercenary soldiers, presumably hired by the pope, defecating in the papal tiara with the caption "The pope is adored as an earthly God" (Scribner 1981:81). In 1528 someone (a Protestant?) published a Papal Bulletin that "enjoined the reading, rereading and dissemination of Luther's works" (Febvre and Martin 1976:290). Warfare via printed posters broke out in several regions. Many of the posters listed banned books or informed everyone about the condemned books.

Several previous commentators had been as severe as Luther in their criticisms of the church but none had rocked the foundation of the Catholic church. Prior to this, the circulation of ideas critical of the Catholic church had been limited to written statements that circulated almost entirely within the church hierarchy and by word of mouth. For example, in the fourteenth century Wycliffe translated the Bible into a vernacular language and denounced the pope. But Wycliffe's ideas did not achieve wide currency. He was branded a heretic, his writings were banned, and his movement came to naught.

The combination of writing in the vernacular and the mass production of books and pamphlets made it impossible for the Catholic church to contain the competing definitions of reality offered by the Protestants. According to the testimony of one contemporary, the home of every peasant had become a school where the Old and New Testaments were studied (Febvre and Martin 1976:294). In 1532, the Catholic church "lamented that nothing seemed to be able to halt the influx of heretical books" (Grendler 1977:75). The spread of

Calvinism as well was accompanied by the establishment of printing presses. Between 1550 and 1564, 527 different books were published in Geneva by forty printing establishments (Anderson 1983:44). Heretical books were readily available in nearly all of the major trade centers and universities by mid–sixteenth century.

Mercenary and temple interests are not always mutually exclusive. Some Protestants published books to support the cause; others to make a profit; still others for both religious and mercenary reasons. Printing and literacy became affiliated in the minds of many with Protestantism. One woodcut shows Luther holding a book while the Devil crouches in a corner in a monk's habit (Scribner 1981:24). Another shows Luther offering a Bible to the pope, who rejects it with an expression of horror (p. 27).

The Jesuit order was established in 1534 to combat the Protestant heresies, reform the Catholic church, and promote the faith among the heathen. Within two decades a score of Jesuit universities were established and most of them operated a printing press. Although the guiding intent was to promote tradition, they contributed to the variety of ideas in circulation.

The Catholic church reactivated the Inquisition in 1252, but the activity was uncommon until the sixteenth century when producers, distributors, and owners of heretical books became the targets (Grendler 1977:34–37). In 1555, the pope "banned the printing and possession of Bibles in any vernacular except with the permission of the Inquisition" and all anonymous works (p. 117). Among the inquisitors, the book became known as the silent heretic.

Several owners of heretical books and a few book merchants were burned at the stake. In 1534, both a printer and a bookseller were burned at the stake (Febvre and Martin 1976:309), and in 1535, six heretics and three large sacks of books they owned were burned in Paris (p. 310). Despite these repressions, the French printers printed as many books in 1535 as in the immediately preceding years. None of the major publishers were targets of the inquisitors (pp. 150–52).

The Inquisition stifled the dissemination of Protestant ideas in some regions, but in other regions it stimulated greater opposition to the Catholic church. Several critics wrote books denouncing the Inquisition. In some cities, the secular authorities resisted the efforts of the inquisitors and protected commercial interests. For example, the authorities of Venice attempted to protect printers and book merchants from the inquisitors (Grendler 1977:37).

In some regions the Catholic authorities were relatively successful in controlling the flow of books, but in other regions Protestant leaders disseminated books denouncing the Catholic hierarchy and called on all to read. The Catholic authorities relied primarily on established procedures to disseminate their definitions of reality while the Protestant leaders made extensive use of the printing industry to promote their own versions. Luther proclaimed that every baptized Christian was a priest and that every person was capable of reading the word of God.

Prior to the Reformation, the Bible was used by the clergy to give authenticity to its utterances. The basic relationship between the clergy and parishioners was that of a speaker and audience. That relationship was largely a consequence of the high cost of reproducing books, low rates of literacy, and the unavailability of Bibles in vernacular languages. The printing press lessened those constraints.

In some circles the belief emerged that it was no longer necessary to attend church rituals to acquire religious instruction and moral dicta. It became common for family and friends to convene in homes to discuss these issues. The belief that the pope was an infallible interpreter of the Bible was diluted. The relationship between readers and the Bible among Protestants was in marked contrast to that of priests interpreting the Bible for parishioners, bishops interpreting for priests, cardinals interpreting for bishops, and the pope interpreting for all.

The hegemony of the Catholic church over religious beliefs and moral issues was shattered, and a host of competing interpretations of the Bible was offered. Several Protestant sects, such as the Quakers, established the policy that each person is capable of deriving moral dicta from the Bible without benefit of clergy. In Protestant groups the content of the Bible often was the topic of discussions, debates, and negotiations. Not all, but a significant percentage of adults acquired access to the information removed from the hierarchical structures that had previously encased its flow.

As Catholics and Protestants competed for the moral hegemony of Europe, the book merchants organized legitimate and clandestine trade routes to distribute books: "Heretical books, made their way everywhere, even into monasteries and seminaries" (Febvre and Martin 1976:316). The leaders of the Reformation turned to printers to diffuse their ideas; owners of presses turned to critics of the Catholic church for manuscripts that would reap a profit.

Some authors who supported the Catholic church complained of difficulties in getting the printers to publish their books (Eisenstein 1979a:354). Criticisms and new definitions of reality sold better than defenses and traditional definitions of reality. As a result, an antagonistic relationship emerged between the Catholic church and the printing industry that endured for centuries. In 1789, the Catholic church disapproved of the French declaration of the rights of man on grounds that it included a declaration proclaiming freedom of the press (p. 355).

During the sixteenth century some Catholics, especially the Jesuits, used the press to advocate traditional definitions of reality, but for several decades, most of their publications were in Latin. Those publications reached only a small number of people and most of those were loyal to the church. In contrast, most Protestant leaders and state authorities of Protestant nations promoted literacy and publication in the vernacular. For example, in Sweden the Lutheran church and the king of Sweden called for all to become literate in the Swedish language and denied marriage to the illiterate. One offshoot of that effort was "Sweden's anomalous achievement of female literacy rates as high as male rates" (Graff 1987:34). The growth of literacy in Sweden reaffirmed Lutheranism, promoted Swedish nationalism, and validated the Swedish monarchy.

The Protestants were most successful in supplanting Catholic churches in regions where the vernacular language was not related to Latin and least successful in regions where a romance language was spoken (Logan 1986:222). There was a larger chasm between the clergy and parishioners in regions where a nonromance language was spoken as the Catholic clergy had a greater monopoly of knowledge. In regions where a romance language was spoken, the dichotomy between clergy and parishioners was less rigid as a minority of the parishioners could also read Latin. The closer integration of the clergy and parishioners where a romantic language was spoken inhibited the spread of Protestantism.

One tangential consequence of the Reformation was the erosion of the significance of paintings, reliefs, and sculptures at religious centers. As Protestantism spread, the emphasis shifted from sacred images to sacred texts. The holy word replaced the holy icon.

Charismatic leaders emerged who offered novel definitions of reality and solicited followers. The alternative definitions of reality offered by the charismatics conflicted with both traditional Catholic definitions and those of Protestant groups that called on all to read, study, and interpret the Bible. In many communities, charismatics and inquisitors competed for moral leadership. In some, the charismatics prevailed, in others the inquisitors prevailed. Some of the charismatics became as repressive and sadistic as the inquisitors.

The Protestant heresies and the Catholic response ignited violent conflict. Warfare stemming from moral confrontations have an ancient past, but the religious wars between Catholics and Protestants achieved a scale beyond any previous violence fueled by religious beliefs. The religious wars had paradoxical consequences. On one hand they intensified religious hatred, but on the other hand they enticed some, especially those active in the bookish culture, to call for religious tolerance.

IMPACT ON OTHER RELIGIONS

If the commercial printing of the Bible contributed to the fragmentation of Christianity, it is reasonable to inquire into the consequences of the commercial printing of religious books on other religions. Until the twentieth century it was uncommon to commercially print the sacred texts of other religions, and so it is difficult to give a definitive answer to the question. We nonetheless can briefly consider the Jewish and Islamic religions.

The printing industry developed as rapidly among the Jews as among the Christians, but in the preprint world the literacy rate among Jews was much higher than among Christians. All Jewish males were obliged to know the Mosaic law (Millard 1972). The literacy rate among Jews increased as printed books became available, but the increase was not as great as it was among Christians. In addition, most Jews were affiliated with markets and many traveled to other regions to trade. They were more aware of alternative definitions of reality than were Christians. The printing press did not significantly modify the information about temple-centered affairs that circulated among the Jews. The printing of sacred texts does not seem to have had major consequences for the Jews.

"It is an axiom of Islam that the Koran is untranslatable, and it is forbidden to attempt a translation" (Boorstin 1985:540). The Koran can be transmitted "only in the original handwritten format used by the Prophet's disciples" (p. 541). A printing press was not established in a Muslim nation until 1727 and that press discontinued operations in 1745. The first printing press in Egypt was established after Napoleon's conquest in 1798. The Koran was printed for the first time in a Muslim nation in 1925 (p. 547).

In most Moslem nations the Koran continues to be reproduced by lithography and in Arabic, and thus the circulation of the Koran in Moslem nations continues to be limited. The prohibitions against printing the Koran and the dictum that it can only be reproduced in Arabic is as if the Bible could only be reproduced by nonmechanical means and only in Latin. If the Catholic hierarchy had been able to enforce an edict that the Bible could not be profaned by translating it into vernacular languages or by printing it, the Reformation might not have occurred.

THE PRESS AND STATE

European state authorities, in contrast to Chinese authorities, were slow to realize that the printing press could be used as an instrument of the state. The printing industry matured most rapidly in cities that were trade centers and the authorities of those cities treated printing much as they treated other commercial enterprises. When the Protestant heresies burst forth, some state authorities aligned themselves with the Protestants, others with the Catholics. Those alignments flowed largely from the religious affiliation of the monarchs. If the monarch was Catholic, Protestant writings were suppressed; if the monarch was Protestant, Catholic writings were suppressed.

A few heads of states used the press to advance their cause. Early in the sixteenth century Emperor Maximilian I "drew up an ambitious programme of 130 books which were to broadcast the glories of the house of Habsburg and especially of Maximilian himself" (Steinberg 1959:45–47). But only two volumes were published before his death.

Some monarchs suppressed books, but as was the case with church edicts, the edicts of state authorities were not adhered to by many printers. A few state authorities encouraged the expansion of the print industry. In 1484 the English Parliament exempted printers and booksellers from restrictions on foreign labor, so that the printing industry would develop more rapidly in England (p. 77).

Printing flourished in those cities that treated printing as just another commercial enterprise. For example, most Dutch cities welcomed printers and many Dutch printers published books in foreign languages. Black market networks routinely distributed books banned by state and temple authorities across national boundaries. The Dutch printers made the evasion of foreign censors into a fine art. A common subterfuge was the faked imprint (p. 189). The imprint on a book published in Holland that advocated the Protestant cause stated that the book had been printed in the Vatican and approved by the pope. Book merchants sometimes bribed inspectors and hid books in bales of clothing, but despite those efforts the inquisitors often were successful in detecting the silent heretic and tens of thousands of books were burned.

Most state authorities who instituted sustained programs of censorship did so at the urging of church authorities. For example, the authorities of Venice attempted to protect book merchants but eventually stationed "an inquisitial representative at the customs house" (Grendler 1977:xx). Banned books, however, continued to flow into Venice.

State authorities were not indifferent to the flow of books in their nations. In 1579 the queen of England became infuriated by a pamphlet that speculated on her marriage and had the hand of the author cut off. Print shops were destroyed, printers were beaten at the behest of state authorities, and a substantial number of authors and printers were jailed or killed by the state authorities. Despite the severe response of state authorities when irritated by authors or printers, their efforts to control printing were only slightly more effective than the efforts of church authorities. It seems that when one printing shop was shut down, two more appeared in its place.

Most printers, transporters, and sellers of the clandestine books eked out a bare existence; others participated in the trade to supplement their regular income. A few publishers that specialized in the clandestine trade, such as Aldine, subsequently became major publishing houses.

In eastern European countries, printing usually was controlled by state authorities. In 1501 the king of Poland granted a monopoly to a printer (Febvre and Martin 1976:202–3) and Ivan the Terrible instituted printing in Russia as "an instrument of the policy of centralization and coercion" (p. 206).

Printing shops were established in the Americas almost as rapidly as permanent communities were. Many of the early American printers produced religious volumes to be used to convert the Indians. In North America, the printing shops were affiliated with universities and a variety of books issued forth; in South America, the printing establishments were tightly controlled by crown and church authorities and "their output was almost exclusively ecclesiastical" (Anderson 1983:61).

INTELLECTUAL FERMENT

Secular books acquired hegemony over sacred ones in the early decades of the sixteenth century. In Paris, in 1501 religious works had a two to one advantage over humanistic ones. By 1549, the ratio had shifted to almost four to one in favor of humanistic books. "Similar inquiries would produce the same results almost everywhere" (Febvre and Martin 1976:264). Simultaneously, secular intellectuals began to replace the clergy as preservers of knowledge. For example, of the estates settled in France between 1480 and 1500 only one lawyer bequeathed a library, whereas twenty-four churchmen did. Between 1551 and 1600, seventy-one lawyers bequeathed a library, whereas only twenty-one churchmen did (p. 263).

Formal education became available to more people as the printing industry expanded. For example, in 1480 England had 34 schools, by 1660 she had 444. Many of the early bestsellers were textbooks. Between 1518 and 1533, one Cologne printer issued "13 Latin grammars, all of which had to be reprinted several times" (Steinberg 1959:100). Whereas previously most children acquired knowledge informally as they associated with adults, with the expansion of formal education an orderly sequence replaced the mosaic exposures of earlier generations. "Textbooks were newly designed to take students in sequence from the most elementary to the most advanced level of a given skill" (Eisenstein 1979a:432). The education of the young moved from an apprentice relationship toward a bureaucratized order.

Intellectual activity both increased and fragmented as the printing industry grew. When Latin was the universal language among scholars, "[t]eachers and students could move from Bologna to Heidelberg, from Heidelberg to Prague, from Prague to Paris and feel at home in the class-room" (Boorstin 1985:489). When scholarly works began to be printed in vernacular languages, however, Latin lost its privileged status and schol-ars not affiliated with the church began to compete with church-affiliated scholars for intellectual leadership.

Prior to the printing press, most intellectual activity focused on reli-gious and moral issues. Afterwards as universities and independent scholars became sources of knowledge, more attention was given to secu-lar issues. New issues came to the fore as widely scattered intellectuals began to publish and send letters to scores of other intellectuals. The intellectual ferment was similar to that which occurred among the Greeks following the spread of phonetic writing into Greek communities.

Intellectual activity also increased in China as printing became com-mon. But in China nearly all such activity was undertaken by people affiliated with the palace. In Europe intellectual activity was more diverse. Some secular scholars raised the ire of temple and state authorities. The universities sometimes served as a haven for those offering novel defini-tions of reality.

In China and the preprint monasteries of Europe, the assessments of superordinates and promotion, not fame, informed the activity of schol-ars. Of course, many scholars, scientists, and engineers of postprint Eu-rope sought approval of temple and state elites as well as fame. But as scholarship based on individuals seeking knowledge increased, scholar-ship dedicated to preserving traditional knowledge declined. The defini-tions offered by the emerging secular scholars were to "the world," not to authorities. The communication of ideas to the public replaced communi-cation to a small select audience as the primary conveyor of truth.

In the script world, instruction at universities usually took the form of a teacher reading from a book to students; with the advent of print that practice was replaced by each scholar reading in isolation. Solitary schol-arship enhanced personal autonomy. Knowledge seeking became more autonomous as secular scholars read, reflected on what they had read, and made observations.

Both reading and writing favor reflective thought. Solitary scholars are more likely to formulate novel definitions than are scholars embedded in enduring asymmetric relationships. Readers and writers who formulate ideas in isolation often think the ideas formulated are a personal accom-plishment. Solitary scholarship promotes individualism.

The printing press stimulated the growth of established universities, facilitated the establishment of new universities, and aided the univer-sities in their competition with the monasteries for the allegiance of intellectuals. The academic efforts of the Jesuits also contributed to the intellectual ferment despite their intentions to promote traditional definitions of reality.

Scholars noted inconsistencies in traditional writings and between the written word and empirical observations, and attempted to remove the inconsistencies. The resolution of inconsistencies between written information and observational experiences was not a novel activity. In the twelfth century Roger Bacon had called for the correction of the calendar, noting that it must be apparent to anyone who would make the observations that Easter was no longer celebrated after the spring equinox. The blossoming of print simply made more nonscholars aware of the inconsistencies between what was written and their experiences and more people became active in efforts to resolve the inconsistencies.

Some began to think of the environment as the "book of nature" and the idea blossomed that if there were inconsistencies between the written word and the book of nature the inconsistencies were to be resolved in favor of nature. Truth became transformed from something to be found in God's word to something to be found in God's work.

The amount of written information about natural phenomena available to scholars dramatically increased during the sixteenth century. For example, the ancient students of plant life had described about six hundred plants and by 1623 the number had grown to six thousand (Eisenstein 1979b:487). Printing facilitated the dissemination of Arabic numbers and computational procedures. A book published in 1585, entitled *The Tenth*, popularized the decimal system of computation in Europe (Boorstin 1985:397). The Catholic church continued to use Roman numerals, but merchants, scientists, and bureaucrats found Arabic numerals more useful.

When scribes reproduced a book they attempted to duplicate the original, but errors were common. Some corruptions were intentional, as when the copyist reproduced a book for his own use and made notes alongside the text. Corruptions occurred in print, but errors were less tolerable. The widespread circulation of printed information encouraged people to be more precise in their presentation of ideas.

Students were freed from the necessity of copying manuscripts and spent more time reading and examining phenomena. Prior to printing, students of astronomy were lucky to have access to one written description of the solar system; by the time of Copernicus most astronomers had access to three different descriptions of the solar system: One description had all celestial bodies circling the Earth; a second had Mercury and Venus circling the sun with all three of those bodies circling the Earth as a unit and the other celestial bodies circling the Earth; and, the third located the Sun in the center of all celestial movement. Copernicus's formulation of a novel theory of planetary movement probably was stimulated by his awareness of three different characterizations of celestial phenomena. Subsequently, Copernicus's theory of planetary movement began to compete with Ptolemy's theory and stimulated additional intellectual activity.

Conflicts between church authorities and scholars occurred, but compared to the wars between the Protestants and Catholics those conflicts were mild. Galileo was confined and a few intellectuals were burned at the stake, but those tragedies touched only a few. As intellectual activity increased, a battle of words between scholars promoting traditional definitions of reality and those promoting novel ones became common. Those in the forefront of this struggle routinely used printing to disseminate their definitions of reality. Copernicus had his theory of planetary movement published in print. Tycho Brahe established a press as part of his research operation. Galileo used the press to propagandize his definitions of reality and ridicule traditional ones.

The ideas of Copernicus, Galileo, and others were read by Catholic scholars, and some became less enthusiastic in their support of traditional definitions of reality. Truth was transformed from something that existed in traditional writing that only a select few had access to, into something to be discovered, discussed, debated, and written about by many.

In a sense the press was responsible for the discovery of the Americas. In the preprint world very few other than those affiliated with *scriptoria* had access to Ptolemy's atlas. A 1482 printing of Ptolemy's atlas showed the Earth as round, displayed the ocean as devoid of land between Europe and China, and described the Earth as smaller than it is. Columbus studied Ptolemy's atlas and concluded the shortest route to China was across the Atlantic Ocean.

Other Europeans had previously sailed to the Americas and returned, but that information did not become part of the collective memory of Europeans. It is possible that Columbus's discovery, too, might have faded from the collective consciousness of Europeans had it not been printed. Columbus's letter to the queen announcing the success of his voyage was printed as an eight-page pamphlet, and became a bestseller (Boorstin 1985:236).

Ironically the press also was responsible for denying the name *Columbia* to the New World. In 1507 a printer published a book about Amerigo Vespucci's "discovery" of a fourth part of the world that contained maps of the New World. When the publisher learned that Amerigo Vespucci was not the discoverer of the New World he attempted to rectify his error, but he was too late. "The printed messages advertising America were already diffused into a thousand places and could not be recalled, and 'America' became indelibly imprinted on maps of the world" (p. 253).

At the beginning of the sixteenth century, European governments encompassed about 10 percent of the world's population. On the eve of World War II, about 70 percent of the world's population was subject to a Western government. European state, market, and temple interests colonized much of the world during the intervening centuries. That colonization was aided and abetted by the print industry.

LUDENIC READING

Some derived pleasure from reading long before print was invented, but after print had been perfected many more read narratives for pleasure. Most ludenic reading, i.e., reading for fun, is for short durations such as glancing at a comic strip. Some sustained instances of ludenic reading involve readers becoming lost in a book (Nell 1988). For most literates, reading requires so much effort that it is difficult for them to get pleasure from reading narratives. Extensive ludenic reading is limited to a minority of readers even in societies that approach universal literacy (ibid.), but nonetheless ludenic reading is a significant new form of activity. Sustained ludenic reading enhances the development of personal discipline more than does sustained listening.

In preprint societies nearly everyone experienced narratives in a collective and expressive context. After print became common, large numbers of people began deriving pleasure from a narrative in isolation with little responsiveness. Ludenic reading almost demands isolation, and responsiveness is untoward when expressed in isolation. Oral performances provided opportunities for people to share pleasant communal experiences and become embedded with one other; printed books provide opportunities for millions to have common experiences, develop personal discipline, and formulate individualistic assessments of complex sets of ideas.

Narratives, whether offered in oral or written form, usually advocate, at least implicitly, a standpoint toward the phenomena described. Printed narratives have the potential of reaching large audiences more or less at the same time. If the proper sentiments are stirred in a large enough audience, authors can stir public controversy. One example of that accomplishment is the novel *Uncle Tom's Cabin*, published in 1850. One and one-half million copies of the book were sold within a year, and public opinion against slavery was crystallized in the northern states.

Some ludenic reading is a search for an alternative reality. The mistreated wife, the would-be-hero, the worker encased in the dull work-a-day world with access to written narratives can, for a time at least, acquire experiences from a book that are more meaningful than those offered by the mundane world. Readers can acquire access to fantastic worlds in solitude. It requires only literacy, access to a book, and a degree of privacy. The reader of a romantic novel can experience romance without risking involvement with another person (Radway 1991:100).

CONCLUSIONS

The printing press was the death knell for orality as a technique for the preservation of information. Between 1500 and 1900, the literacy rates of Western European nations grew from an estimated 5 to over 90 percent. Poems continued to be written but they were composed in writing, not orally, and they were no longer composed to preserve information, but to entertain.

Writing had been used for millennia to preserve and disseminate large quantities of information, but with few exception the books reproduced prior to print circulated only among small elite populations. Printed books distributed in markets made available complex sets of ideas in artifactual form to large numbers of people, many of whom were not embedded in either a temple or palace hierarchy. Many were enticed to examine and reflect on book knowledge. In regions where printing was a state operation or controlled by state and temple authorities, there also was an increase in the percentage exposed to complex ideas, but the increase was much less. Furthermore, when printing was a state or temple operation, novel definitions of reality were much less likely to achieve wide circulation.

The printing industry matured in the marketplace in Europe. When the temple and state authorities responded, they attempted to control what was printed, but so many printing presses were in operation that it was difficult to constrain them. Nearly all who had the opportunity to publish their ideas in China were ensconced in the palace-centered state structure; consequently novel definitions of reality were less pervasive in China than in Europe.

When printing was controlled by state authorities a new form of totalitarianism that rested on the dissemination of a monolithic definition of reality via print emerged. The authorities of China, Nazi Germany, the USSR, and many other nations tightly controlled the production and distribution of books. In such nations, printed information was not associated with social turmoil.

The state authorities of phonetic societies have found it difficult to control the black market distribution of books. In Europe when state authorities took notice of printing, "there came into being a mass of draconian legislation, precise enough to meet all eventualities, yet generally ignored" (Febvre and Martin 1976:311). Some modern state authorities have been more successful in controlling the flow of ideas via books than were the European state authorities of the sixteenth and seventeenth century. But no modern state has completely stemmed the distribution of heretical ideas via books. Even when exceedingly repressive measures have been instituted, some subversive books continue to be distributed via black markets.

In most regions of the world, books became the dominant repositories of knowledge by the end of the nineteenth century and knowledge became something to be preserved as an enduring artifact. Those imbued in the world of print consult books, not people, when searching for knowledge. When a multitude of books are available, literate people may consult a multitude of sources without becoming accountable to anyone. Print creates conditions wherein people can know of others and be immersed in disembodied ideas without any involvement with other people. Consciousness derived from print is emotionally impoverished, but more complex than that derived from orality.

CHAPTER

9

Newspapers and State Structures

Some printed information keeps an indefinite relevance, while other printed information is relevant for only a short time. In general, books preserve information of enduring relevancy, and newspapers disseminate information of fleeting relevancy. One consults a book, for example, to learn the history of a nation, and a newspaper to learn what happened in the national capital yesterday. Books preserve and dispense timeless information; newspapers dispense timely information.

News—information about recent events—was disseminated long before newspapers were invented. In nomadic societies, for instance, when one person noted the approach of a stranger, he typically informed others, and people continue to share news with family, neighbors, and coworkers via the spoken word. Town criers disseminated news in Mesopotamia during the third millennium before the Christian era (Finkelstein 1979:52) and continued to do so in many regions of the world until this century.

Written news probably first appeared in ancient China about 200 B.C., when state officials posted notices on walls. At the time of Julius Caesar, Roman officials posted bulletins to inform citizens of recent developments. In preprint Europe, "News about local and foreign affairs, real estate transactions, and other mundane matters was regularly announced in churches" (Eisenstein 1979a:131).

Today, newspapers are taken for granted. Their distinctiveness from books in earlier times is indicated by the reaction of a young Japanese scholar when he first encountered newspapers in the nineteenth century. The scholar had been assigned the task of translating the contents of Dutch newspapers into Japanese for a warlord. He was startled to learn that Westerners printed each day a thing called a newspaper that informs men of both foreign and domestic events. He was also puzzled by the range of topics addressed and asked the Dutch sea captain, from whom he obtained the newspapers, how one author could address so many topics (Huffman 1980:16). The scholar subsequently established one of the first Japanese newspapers.

EARLY NEWSPAPERS

Early newspapers were called news books and disseminated information about the activities of the elite and economic affairs. One newspaper in 1623 reported on a meeting of the emperor and princes, the preparations of a king for a trip, market conditions in foreign countries, and the election of a pope. In a similar manner, contemporary newspapers report on the meetings of heads of state, inflation rates, the affairs of the elite, international conflicts, and domestic disturbances.

Irregular publications that were precursors of newspapers appeared in Venice in the 1560s and reported the arrivals of ships, items for sale at the markets, and foreign affairs. They were sold on the streets for a *gazeta*, a coin of small value; hence the name gazette for many newspapers. A newspaper appeared in Antwerp in 1605 (Altschull 1984:10), and "[w]ithin a decade, newspapers were founded in Frankfort, Berlin, Hamburg, Basel, Vienna, and Amsterdam" (Brasch and Ulloth 1986:37). The early newspapers only had two or four pages and usually only a few hundred copies of each edition were printed.

Whereas European authorities had been slow to react to the book-publishing industry, most of them responded quickly to newspapers. The authorities were particularly disturbed by published descriptions of domestic affairs, especially accounts of foibles of the domestic elite. An English king acquired the honor of becoming the first monarch to suppress newspapers when in 1632, only ten years after the first newspaper was published in England, he banned all news books (Steinberg 1959:174). The decree was lifted six years later, but permission was only granted to print foreign news (Small 1982:55). The Catholic church's control of newspaper publishers was even less effective than its control of book publishers, although "[t]wo popes went so far as to ban the writing of avisi (news) on pain of branding or sentence to the galleys" (Altschull 1984:8).

Nearly all early newspapers were commercial enterprises. A few state-operated newspapers were established in the seventeenth and eighteenth century, but they were of minimal significance. In 1631, the French government sponsored a newspaper but it was discontinued on the death of the king, who had initiated the undertaking (Steinberg 1959:174). No state-operated newspaper had a massive circulation until the twentieth century.

In 1643 in response to the sustained conflict between the Puritans and the royalist Catholics that was fanned by newspapers and pamphlets, the British Parliament attempted to prohibit the publishing of inflammatory statements. In 1644, Milton, an advocate of the Puritan cause, rationalized the concept of a marketplace of ideas on the grounds that a free flow of information led to truth. He called for unlicensed liberty for printers, which struck a responsive chord among publishers and readers.

When the Puritans became victorious, Milton became the state censor and was as vigorous in his censorship of newspapers as had been his predecessors. Nonetheless, the revolutions that overthrew the English monarch and established Cromwell as dictator and that later replaced Cromwell with a new monarch rendered state control of newspapers somewhat more problematic in Britain than in most other nations. Throughout the seventeenth century there was an erratic growth of newspapers in western Europe. Calls for removal of state restrictions on the press became more common, and, not coincidentally, the term *democracy* entered vernacular English in the seventeenth century (Small 1982:35).

When confrontations between state authorities and newspaper publishers occurred, the publishers usually elicited more sympathy than the state authorities. For example, in 1703 when Daniel Defoe was sentenced to the pillory for publishing a statement critical of state authorities, he was treated as a hero and sympathizers drank toasts in his honor while he was in the pillory and brought flowers to decorate it (Ingelbart 1987:74). Six decades later, in 1763, a critic of the king was fined for seditious publication, and "the public contributed more than 10 times what was needed to pay the 1,000 pound fine" (p. 113).

A few printers of newspapers received the death sentence for defying authorities and several were imprisoned. When presses were destroyed, however, new ones usually took their places. From the mid–seventeenth through the eighteenth century there was almost constant conflict in Britain between state authorities and newspaper publishers, but the newspaper publishers in France and Germany "accepted the directives of the crown without apparent protest" (Altschull 1984:14).

In a few regions, newspaper publishers were not severely censored. In the mid–seventeenth century, Dutch newspapers began to report on both domestic and foreign state affairs and "transformed state events into public events by making the mere reader privy to the doings of the mighty" (Schama 1987:268). Printed news of the doings of the mighty became an exciting novelty in a few nations. Many newspapers printed criticism of foreign authorities and domestic authorities seldom attempted to suppress such items. For example, in 1662 the English authorities formally complained that a newsmonger of Utrecht had taken the insolent liberty of "speaking of His Majesty and Government" (ibid.) and called on the Dutch authorities to correct the situation. The protest fell on deaf ears.

Nearly all early newspapers were fiercely partisan and often challenged the accounts provided by competing newspapers. "Indeed 'liar' was probably the most used epithet in all their very rich language of abuse" (Small 1982:62). Early newspapers competed with one another to establish their definitions of reality as the authentic ones. Their competition with one another contributed to the emergence of a concern with freedom of the press in the political discourse of the seventeenth century.

The rise of the newspaper industry swelled the ranks of those who adopted a reflective standpoint toward authorities. The term *public opinion* first emerged in the early eighteenth century, and by the 1780s the term had become current in the public discourse of state affairs (Cmiel 1990:25). Whereas the information provided in books tended to stimulate abstract discussions of state structures and historical figures, the information provided by newspapers stimulated agitated discussions of concrete events and the merits of contemporary state authorities. The information provided by newspapers rendered state authorities more profane and generated reflective assessments of them.

NORTH AMERICAN NEWSPAPERS

Many Europeans migrated to North American to avoid repression by state authorities for their heretical ideas. All the colonies formed assemblies that enacted legislation, but the assemblies were, theoretically at least, subordinate to Parliament and the king. Despite their supposed allegiance to England, most of the assemblies formulated policies that implicitly presumed their colony was an autonomous unit. Many members of the colonial assemblies were aware of alternative state structures. "In studying the laws which were promulgated at this early period, it is impossible not to be struck by the remarkable acquaintance with the science of government which they display" (de Tocqueville [1835] 1956:46).

Large quantities of books were imported from Europe and the literacy rate, especially of the northern colonies, was higher than that of any European nation. In some colonies as many as 90 percent of the adult males were literate (Hart 1950:8). Despite the rigid control of printers in England, by the mid–seventeenth century printing presses were operating in Cambridge, Boston, and Philadelphia "without resistance by the Crown" (Postman 1985:36). The licensing of printers was never effectively imposed in the colonies, and newspaper publishing acquired primacy over book publishing.

The growth of newspapers in the colonies was enhanced by the interests of the colonists in news from Europe. Whenever a ship arrived in a city where a newspaper was published, news of its arrival, its cargo, and information the crew had about European affairs was printed. Many of the printer-journalists of North America also were postmasters. The first colonial newspaper was founded in 1704 by the postmaster of Boston. Ben Franklin became both the leading newspaper publisher and the colonial postmaster.

From the beginning, colonial newspapers published criticisms of authorities. In 1723, the printer of a Boston paper was arrested for publishing criticisms of the king, but so many citizens rallied to his support that he was never brought to trial. Another newspaper publisher spent a month in jail for criticizing the authorities' lack of action against pirates. Whenever the crown-appointed authorities and colonial assemblies came into conflict, the confrontations usually were reported in the newspapers. Newspaper publishers nearly always sided with the assemblies. Reciprocally, when the crown-appointed authorities attempted to control the newspapers, members of the assemblies usually defended the publishers. For example, in 1747 the New York assembly resolved that efforts to prevent publication of its proceedings "is a violation of the rights and liberties of the people" (ibid.). The colonial newspapers and assemblies supported one another; in contrast, the English House of Commons did not allow newspapers to report on its proceedings until 1771 (Wiggins 1980:366).

In 1735, Zenger was charged with "Scandalous, Virulent, and Seditious Reflections" about the governor (Small 1982:78). One indication of the difficulties confronting the colonial authorities is that Zenger refused bail and continued to operate his newspaper through "a hole in the prison" while awaiting trial. The jury saw fit to acquit him on the grounds that what he had printed was true, and upon his acquittal the community leaders retired with Zenger to a tavern to celebrate. Zenger's acquittal was in direct violation of the legal code, since at that time the truthfulness of statements critical of the king had no legal standing. The prosecuting authority noted that truthful seditious statements were more damning than false ones and therefore if the jury thought his statements to be true that increased his guilt. The jury disregarded that appeal. The truthfulness of printed statements was not legally recognized as a defense against sedition in Britain until 1843.

Publishers were emboldened by Zenger's acquittal. No other colonial printer was brought to trial, although many continued to publish statements critical of the authorities. The risk of publishing criticisms of officials was often lessened by publishing anonymous letters, many of which were written by the publishers themselves. In 1755 the *Boston Gazette* claimed the "right to expose abuses of power" (Smith 1988:7), and Franklin advocated complete freedom of the press when addressing "the propriety of public measures and political opinion" (p. 11). In the decades preceding the revolutionary war, the issue of "freedom of the press" acquired wide currency.

In 1775, forty-two newspapers were being published in the colonies (Botein 1980:417), and probably one in twenty families received a weekly newspaper in 1765 (Kielbowicz 1989:20). The newspapers were filled with arguments and counterarguments about affairs of state (Bailyn 1967:1). Edicts of the crown were ridiculed in the newspaper. Criticisms of crown-appointed governors and magistrates were sometimes accompanied by statements that condemned all forms of authority.

Newspapers also disseminated descriptions of the dastardly acts of the British authorities. For example, in May 1768 in a confrontation with a crowd, a British regiment killed several people and tracked down and shot a youth. Newspapers published lurid descriptions of the affair. One critic noted that newspapers were stuffed "with the most inflammatory pieces they procure, and excluded everything that tended to cool the minds of the people" (Botein 1980:12–13). When the governor of Massachusetts in 1768 called on the house of representatives to charge the *Boston Gazette* with libel, the assembly refused to act on the grounds that "the liberty of the press is the great bulwark of the liberty of the people" (Buel 1980:59).

None of the newspapers had a massive circulation, since a manual press could only print about two thousand sheets in a ten-hour day. But the information provided by newspapers achieved a far greater circulation than the number of papers sold. Newspapers were commonly tacked on the walls of stores and public buildings. State-related news was discussed in taverns, in coffeehouses, and on the street. Newspaper circulation increased as antagonisms toward Britain intensified.

In 1765, Parliament enacted the Stamp Act to control newspapers but instead of having the intended effect it unified the publishers and began a era dominated by discussion of freedom of the press (Mott 1962:63). A similar tax had been imposed in Britain for over half a century and "was not generally regarded as intolerable" (Botein 1980:23). In contrast, the colonial newspapers gave "full accounts of the proceedings of all the colonial legislatures, town meetings, and other bodies which protested against the Act" (Mott 1963:72) and a "flurry of pamphleteering" protesting the act followed (Bailyn 1967:4). The *Maryland Gazette* ran the headline that the *Gazette* had died and published a skull and crossbones cartoon when it reported passage of the Stamp Act.

The antagonisms inherent in the colonial relationship and the difficulty of maintaining surveillance of colonial publishers made it impossible for the British authorities to enforce the Stamp Act. The Committee of Correspondence was established in 1772 to facilitate the distribution of information in the colonies. The following year, loyalist postmasters began destroying seditious mail. The revolutionists took over the operation of the postal system six months before the Declaration of Independence (Kielbowicz 1989:22).

Newspapers popularized the slogan No Taxation without Representation. Thomas Paine's *Common Sense* went through twenty-five editions, and it is estimated that one in five citizens possessed a copy (Cmiel 1990:50). "Within the year 1776 most of the literate men in America had read it" (Mott 1963:91). Paine popularized the terms *freedom, liberty,* and *independence.* "Even the word revolution was transformed in his writings, from a term derived from the motion of the planets implying a cyclical view of history to one signifying vast and non-reversible social and political change" (Foner 1976:15). Concurrently, assemblies and town meetings condemned English authorities. For example, in 1774 a thousand residents of Farmington, Connecticut, resolved that the British authorities were instigated by the devil and composed of pimps and parasites (Bailyn 1967:125).

Prior to, during, and after the revolutionary war, state structures other than the monarchy were widely discussed, written about, and negotiated, and the idea "that freedom of discussion is the bulwark of liberty" became widely accepted (Nerone 1990:13). Prior to the American Revolution, ideas of that sort were discussed only in the European bookish culture, but among literate Americans they achieved almost universal currency.

The newspapers provided a vocabulary that justified rebellion not merely against a particular monarchy, but against monarchy itself as a form of government that required a new type of state. In 1815, John Adams offered the assessment that "the Revolution was in the minds of the people, and this was effected, from 1760 to 1775" through newspapers and pamphlets (as quoted in Bailyn 1967:1).

The American Revolution was the first colonial rebellion populated by people informed with a vocabulary provided by newspapers that rationalized opposition to state authorities. Some pamphlets and newspapers that justified monarchy and the English monarch also were circulated, but they were fewer and commanded far less attention. In 1775, a mob supporting the revolution smashed a press that had published criticisms of the rebellion (Ingelbart 1987:107). Shortly thereafter the publisher of the paper returned to England. Tories were, for the most part, a silent minority.

Long before the American Revolution, alternative forms of statehood had been written about and discussed, and a concern with freedom to express assessments of state affairs was already current in some circles. The Greeks of the classical age wrote about and debated forms of state structures and several European scholars had published analytical statements in books prior to the eighteenth century. Members of Parliament had claimed the right to freedom of speech in the fifteenth century (Mackenzie 1959:34), and in 1688 a law was enacted that gave members of Parliament freedom of speech while in session (Brasch and Ulloth 1986:46). However, only a very small percentage of European citizens concerned themselves with forms of state structures or freedom of speech.

In Europe, most printed information about state structures appeared in books. In America, information about state structures appeared in newspapers as well as in books. The idea that an unfettered press provided a free marketplace of ideas from which truth would emerge became widely accepted among literate Americans. During the prerevolutionary period, European authorities routinely condemned licentious newspapers and European intellectuals offered rationalizations for monarchy. For example, in 1787 George IV proclaimed that licentious publications dispensed poison and had to be suppressed, and Blackstone, a British scholar, published the claim that in all forms of government there must be "a supreme, irresistible, absolute, uncontrolled authority in which the rights of sovereignty reside" (as quoted by Schwartz 1988:5).

The issue of how the governed should select their governors also was widely discussed and written about. Some advocated procedures that would lead to the selection of state authorities who would speak for national interests; others advocated the selection of governmental officials who would represent local interests. This issue became a bone of contention for the Federalists and anti-Federalists.

From Zenger's trial until after 1789, no state structure capable of effectively controlling what was published in newspapers was in place. A de facto unfettered press had become almost traditional by the time a state structure was put in place in America and even then the Federalists resisted the creation of a strong central government.

European publishers, writers, and editors on occasion challenged the legitimacy of state control of the press, but the idea of a weak central government with no control of newspapers by the state was seldom advanced. In contrast, in the United States many equated control of the press by state authorities with tyranny. Prior to ratification of the First Amendment, ten state constitutions "explicitly provided that the press should be free" (Buel 1980:60). The ratification of the First Amendment in 1791 formalized the de facto freedom of the press that had prevailed for some time.

Constraints on newspapers abated more rapidly in North America due to the peculiarities of the relationship between the colonies and Britain and the historical coincidence of a successful colonial rebellion while the newspaper industry was maturing. "The press enjoyed more freedom in post-revolutionary America than it had ever before known anywhere in the world. Nowhere else in the world could a treaty with a foreign power have been discussed before ratification as was the Jay treaty in America" (Mott 1963:143). The post–Revolutionary War newspapers almost routinely published mud-slinging invectives. Washington was depicted as the corrupter of a young woman and John Hancock as the father to a brood of illegitimate children (Bailyn 1967:17).

Some state authorities called for censorship of newspapers. When Adams became president, antisedition laws were enacted that made it a crime to publish any false, scandalous, or malicious information about the government. A few printers and authors were jailed. The legislatures of Virginia and Kentucky resolved that the Sedition Act was unconstitutional. Even the Federalist senators supporting the Sedition Act "disapproved any intention to impose prior restraints" on newspapers (Wiggins 1980:368). The efforts of the Federalists to silence criticisms under the antisedition laws "engendered sympathy for the oppressed Republicans and hastened the Federalists' decline" (Kielbowicz 1989:41).

When Jefferson succeeded Adams, the Sedition Act was repealed and those jailed for violating the act were released from prison. After the Jeffersonians acquired control of the federal government, the Federalist newspapers were subjected to scathing attacks. A mob attacked one printing shop of a Federalist newspaper and forced it to discontinue publication, but the state authorities took no actions against Federalist newspapers (Buel 1980:95).

Next to issues that centered on governing by consent and the accountability of state officials to citizens, press-state relationships were given wider currency by newspapers than any other state-related topic. Jefferson's assertion, "The public judgment will correct false reasoning and opinions, on a full hearing of all parties," reflected a belief that had almost universal currency in the publishing industry.

The adoption of the First Amendment transformed the formal relationship between the publishing industry and the state. Previously conflicts between publishers and the state had taken the form of publishers antagonizing state authorities and suffering the consequences. The First Amendment charged state authorities with sheltering those who disseminated information that stirred the wrath of state authorities. That obligation has not always been fulfilled, but it provided publishers, editors, and journalists an official rationalization that they used to justify the distribution of criticisms of state authorities.

The First Amendment was the first state document to conceptualize speech as distinct from action. That dichotomy was not made in preprint societies, nor is it made in totalitarian societies. It was only after large quantities of information were circulated via print that information could be conceptualized as distinct from action. The dichotomy between speech and action continues to be negotiated in the courts. For example, a 1989 ruling of the U.S. Supreme Court asserted that the burning of the flag in protest of state activities is a form of speech protected by the First Amendment.

There was movement toward less state control of the press in western European nations during the eighteenth century, as a series of confrontations between publishers and state authorities unfolded. The confrontations swelled the ranks of those who adopted a reflective standpoint toward state structures. In some nations, calls for freedom of the press were effectively stifled, but in nations with high literacy rates it became increasingly difficult for state authorities to suppress newspapers.

In the nineteenth century, reports of scandals and criticisms of state officials were more common in American newspapers than in the newspapers of any other nation. In the 1830s de Tocqueville noted that American journalists had a vulgar turn of mind and relished disclosing the vices of public figures. Some state authorities called for the newspaper industry to be brought under control, but support for a free press was so widespread that it was impossible for state authorities to effectively control newspapers.

LEGISLATURE-CENTERED STATES

Apologists for totalitarian state structures have argued that state order requires investing one person with absolute power. Those who drafted the U.S. Constitution rejected that conception of state structures. They feared tyrants, demagogues (charismatics), and monarchs. Several of them wrote statements that dissected the structures of ancient and contemporary states. Those in the forefront of the revolution were especially intrigued by how ancient Rome had been transformed from a republic to dictatorship.

One of the major issues was how to establish a state structure deemed sufficiently powerful to maintain social order and prevent foreign invasions yet one that would not transform itself into a tyranny or monarchy. In addition to the federal constitution in 1776 eight states drafted and adopted constitutions. Newspapers and pamphlets disseminated information about these constitutions and published many assessments of them. The federal constitution came to be judged inadequate and a constitutional convention was convened. The delegates at the constitutional convention were acting within a context where a substantial percentage of the citizens had informed opinions about state structures.

A major division was between advocates of an energetic government and those who feared it. Nearly all of the delegates agreed that men lusted for power. As the Federalists and anti-Federalists confronted one another, they probed the issues: How can the passion for power be contained and how do citizens give consent to state authorities? As negotiations unfolded, procedures for controlling the passion for power and the desired link between the people and the state took clearer form (Ketcham 1986:310). A consensus was achieved that the state should be composed of three branches, with each branch checking the power of the other two. One branch was to be a legislature populated by representatives accountable to constituencies. The delegates transformed the slogan No Taxation without Representation into a principle of government.

The delegates established the House of Representatives as the legislative center, with each representative directly authorized by and accountable to specific constituencies. The representatives were empowered to formulate policies, such as taxation rates, but they were not empowered to issue orders. Constituencies were empowered to instruct their representatives. A state structure was created wherein elected representatives were continually accountable to their constituencies. Paraphrasing James Madison, a state structure was put in place wherein ambition was checked by ambition, not by parchment.

Many other legislature-centered states had been put in place, but the more ancient legislatures were not populated with representatives accountable to particular constituencies. For example, classical Athens was governed by a legislature accountable to the citizens of Athens, but Athenian legislators were not clearly differentiated from other citizens. Other citizens had the prerogative of participating in the legislature's deliberation, and they sometimes did so. The actions of the Athenian legislature were regarded as the actions of the city. Members of the assemblies of Greece were not accountable to a specific district, population, or organization; nor were they authorized to act behalf of others.

In a similar manner, some of the commercial cities of Europe such as Venice, Florence, and Geneva had state structures wherein assemblies composed of citizens were the focal point. "In theory, offices were to be rotated among all citizens of all factions" (Schwartz 1988:112), and each citizen when fulfilling his obligation "was to act as if he embodied the whole" (p. 124).

Authorization and accountability were viable features of those state structures, but the authorization and accountability were not formally linked to particular populations. Instead, all legislators were diffusely accountable to all citizens. In those republics when accountability was instituted, it "occurred mainly in the extreme: citizens would be exiled from the city" (p. 118). The Athenians banned from the city for ten years those citizens thought to have acted inappropriately. Other nations had state structures that included assemblies populated by persons who implicitly represented regions, cities, or special interests. For example, the members of the House of Lords and House of Commons of England implicitly represented constituencies. But their constituencies did not formally authorize those who met in Parliament to act in their behalf nor were there formal procedures for citizens to call members of those assemblies into account.

Prior to the American achievements, the term *representative*, a Latin derivative, referenced a relationship between a spokesperson and those spoken for. In ancient Rome only citizens could speak in court and those who advocated the interests of other family members and slaves were called representatives. Such people were not explicitly accountable to those they spoke for in court. In contrast, when the U.S. House of Representatives was institutionalized, its members were authorized by constituents to "re" "present" the interests of the constituents and were accountable to them.

The representative relationship hammered out by the delegates at the Constitutional Convention presumes that constituencies have the ability to formulate collective interests and to instruct representatives. Those abilities in turn presume that constituents are informed about state affairs. Constituencies are more likely to have the necessary information about state affairs to maintain a representative relationships when they have sources of information, such as commercial newspapers, about the actions of their representatives.

Representatives acquire more information about state affairs than do constituents. Most quickly become aware that the intensity of their accountability can be lessened if the constituents are uninformed or misinformed about their activities. Representatives often manipulate the flow of information to lessen their accountability. When representatives control the flow of information about their activities, they often offer definitions of reality that promote self-interest and fail to advance collective interests (Katovich, Weiland, and Couch 1981; Weiland and Couch 1986). Representative democracies have only endured when information about the activities of state officials has been distributed as a commercial commodity.

Viable representation can be maintained in local communities when constituents directly monitor their representatives and obtain information about their activities via the spoken word. Those procedures for obtaining information about state affairs are not sufficient for maintaining democratic structures that encompass populations of millions.

In authoritarian states, citizens are under the surveillance of officials and are accountable to them; conversely, officials are only incidentally monitored by and accountable to citizens. In representative democracies, representatives are intensely accountable to their constituencies; reciprocally, constituencies are only minimally accountable to their representatives.

The participants in the social engineering that culminated in the institutionalization of the first representative democracy were aware that they were constructing a novel state structure, and they were pleased with their achievement: "I am bold to assert that it is the best form of government which has ever been offered to the world" (James Wilson, as quoted in Ketcham 1986:188). Most intellectuals of the day and age doubted that representative democracy could maintain civilization (Burke [1790] 1968). Some continue to doubt the viability of representative democracy.

THE DIFFUSION OF REPRESENTATIVE DEMOCRACY

The success of the American Revolution enticed many Europeans to advocate an unconstrained press and representative democracy. A series of violent confrontations between the advocates of representative democracy and defenders of traditional state structures unfolded.

In some instances, as in Britain, Sweden, and Holland, there was an incremental movement toward representative democracy and freedom of the press. During the eighteenth century, English juries became "more unwilling to convict for seditious libel" (Buel 1980:68), and the press acquired greater freedoms. In other nations, such as France, Spain, and several Latin American nations, civil wars erupted that culminated in a series of cataclysmic transformations from palace-centered state structures to legislature-centered ones and back to palace-centered ones. Several European nations instituted state structures that were composite palace and legislature-centered structures and several dictatorships were established.

Newspapers and pamphlets were widely used to criticize traditional state structures and advocate democracy. For example, the last decade of the French monarchy "witnessed a proliferation of ephemeral literature of all kinds—newspapers, literary journals, brochures and pamphlets, printed ballads and poems" that offered critical assessments of the king, monarchy, and the state structure (Schama 1989:176). One pamphlet offered an account of King Louis's use of one thousand girls over a ten-year period (Darnton 1982:145). The publications, in effect, demystified monarchy.

After discontent became widespread, the king of France convened the Assembly of Notables to shore up the legitimacy of his rule, but some notables criticized the king. "The more vocal their (the notables') complaints, the more enthusiastically they were applauded in pamphlets and broadsides. The lapdogs of the government turned into terriers of the people" (Schama 1989:260). The revolutionists ultimately replaced the king with the assembly as the focal point of the state structure of France.

When the king was displaced in 1789, the press was granted freedom. The number of newspapers published in Paris exploded from 4 in 1788 to 355 in 1790. The glory days of newspapers in France, however, were short-lived. Many of the newspapers published news that was as critical of the assembly as earlier news had been of the king. In 1791 the French Assembly banned "voluntary slander against the probity of civil servants" and editors were imprisoned and their presses smashed (Altschull 1984:14–15). The number of newspapers in France precipitously declined immediately. When Napoleon came to power he asserted, "Newspapers say only what I wish" (Ingelbart 1987:179), and the newspapers of France became organs of the state. Nonetheless, during their brief time in the sun, newspapers disseminated information that rendered monarchy problematic in France.

One factor that contributed to the different trajectories of newspapers in North America and France was the absence of a viable state structure in North America during the revolutionary war. The state structure of France experienced several violent convolutions during and following the French Revolution, but functionaries were continually on hand to enforce the edicts of whoever had control of the state structure. In contrast, very few functionaries were available to the state authorities in the United States when a legislature-centered state structure was being put in place. Consequently, de facto freedom of the press prevailed in North America for several decades before a state structure capable of effective censorship was in place. During that period the belief that the dissemination of competing definitions of reality contributes to communal welfare became more widely accepted in North America than in France. In France, freedom of the press burst forth suddenly and prevailed only for a short time before the French assembly reinstituted censorship.

Holland also moved toward representative democracy in the 1780s. Preceding and during the turmoil, Dutch newspapers published scandals and criticisms of state authorities. The Dutch press "denounced Prince William V of Orange as a drunken imbecile" and criticized aristocrats for "attempting to preserve systems of nepotism and oligarchy (and as) a Gothic monstrosity" (Schama 1989:249–50). Efforts by the authorities to control the press resulted in "editors and publishers becoming overnight popular heroes" (p. 249).

Confrontations between the newspaper industry and state authorities continued throughout the nineteenth century in western Europe. The authorities of many states suppressed the flow of information about the American and French revolutions. For example, in 1790 Spain forbade all books printed in France and the publication of information about the

French Revolution. In 1826, after the French had installed a constitutional monarchy, the king called for state control of newspapers. That led to "the most extensive public debate the world has ever experienced on the nature of the press and its role. . . . Every day fascinated crowds gathered outside the Chamber to jeer the Parliamentarians and to vent their collective anger against the law that would have imposed a stamp tax on all printed matter" (Altschull 1984:31–32). The king abdicated.

The newspaper industry and representative democracy were slow to diffuse to non-Western nations. The first newspaper published in China was established by foreign commercial interests in Hong Kong in 1858 (Cheng 1970:48). Shortly thereafter a state-operated newspaper that reported on state appointments, travels of the emperor, and memorials was established in Peking. It was distributed to state officials in Peking and provincial authorities.

The first commercial newspaper of Japan was published in 1868 and within a few months fifteen more newspapers were established (Huffman 1980:49). One claimed a circulation of fifteen hundred and "aimed to make private secret matters widely known" (p. 47). Another editor announced that his newspaper could be read by all, including women and children (p. 50). The state officials reacted quickly: in less than a year Japanese editors were jailed for offending state authorities (p. 57) and all but a few newspapers discontinued operations. Other newspapers were established, but those that endured were controlled by state officials or relatives of officials (p. 81). In 1875, a law was enacted that made it a crime for newspapers to publish information that would "confuse the sense of duty of citizens" (p. 104).

Despite state suppression, Japanese newspapers acquired considerable political influence in the 1870s. The ability of the Japanese state to control newspapers waned (p. 82), and calls for the establishment of a popular assembly proliferated. In 1883, the government changed its tactics and began publishing a newspaper (p. 156). That partially defused agitation for representative democracy, but a Japanese parliament was established in 1889. By 1920, political parties were a significant feature of the Japanese state (McNelly 1987:102). The economic depression of the 1930s led to the decline of political parties and the military emerged as the dominant factor in the Japanese state structure.

State-operated newspapers appeared in the Muslim world in the nineteenth century but were distributed only to state officials (Mardin 1979:405). A Briton established the first commercial Arabic language newspaper in 1840 (ibid.); a second commercial Arabic language newspaper was established twenty years later. The offices of the commercial newspapers became gathering places for opponents of the ruling elite (p. 415) and state censorship of commercial newspapers was imposed in 1864.

The principle of freedom of the press has been institutionalized only in nations with high literacy rates and commercial newspapers. Rebellions were a common phenomenon long before commercial newspapers. Most rebellions were localized and state authorities could usually marshal sufficient forces to crush them. The printing of information about current state affairs and descriptions of alternative state structures sometimes generated antagonisms on such a scale that state authorities had difficulty in suppressing rebellions.

An increase in newspaper reports on state affairs accompanied all installations of representative democracy and all returns to authoritarian state structures were accompanied by state authorities instituting censorship of newspapers. The decline of monarchies and dictatorships where edicts are the order of the day and the rise of legislature-centered states where negotiations are the procedure whereby policies are formulated was contingent on a newspaper industry relatively free from state control. The relationship between the state and the newspaper industry and the preferred form of state structure continue to be controversial issues in many nations.

MASS DISTRIBUTION

In the United States, newspaper publishers were not required to obtain a license, post a security, or pay a stamp tax, as was the case in most nations. "The consequence is, that nothing is easier than to set up a newspaper. . . . The number of periodical and semi-periodical publications in the United States is almost incredibly large" (de Tocqueville [1835] 1956:93). Nonetheless, until mid–nineteenth century the newspaper industry, even in the United States, was a modest one.

The first steam-powered rotary press became operational in 1811. Within fifteen years the technique had been improved so that twenty-five hundred pages could be printed in an hour. Newspapers quickly achieved a new plateau. The steam-powered rotary press allowed for the transformation of newspapers from *a* source of news to *the* source of news. On September 3, 1833, the *New York Sun* became available so cheaply that it achieved a circulation of thirty thousand within a year (Brasch and Ulloth 1986:121). In 1855, a British newspaper achieved a circulation of 350,000 (Small 1982:93). "In the late 1880's in America the combined circulation of daily newspapers for the first time surpassed the total of urban households" (p. 101).

One of the founders of the penny press commented: "Books have had their day—the theaters have had their day—the temple of religion has had its day. A newspaper can be made to take the lead of all those in the great movements of human thought and human civilization" (Bennett, as quoted in Small 1982:94). One editor of a mass circulation British newspaper asserted, "[M]ethinks the editor should be the very Moses of our nineteenth century . . . the Captain of our Exodus into the Cannan of a truer social order" (Jane Lowell, as quoted in Boston 1988:94). The claims were a little grandiose, but in representative democracies as newspapers achieved massive circulation, editors and reporters competed with state officials and political candidates to set national agendas.

As commercial newspapers became household items in representative democracies, newspaper editors and reporters established themselves as both *the* source of information about state affairs and *the* expression of popular assessments of state officials. The idea of the press as "the fourth estate of the realm" was first formulated in 1828 (Small 1982:77) and in all modern representative democracies newspaper people regard themselves as the fourth branch of government (Carter 1959).

Many reporters and editors have "draped around their shoulders the mantle of defender of freedom and scourge of abusive power" (Altschull 1984:33) and some of them have taken their self-assigned responsibilities seriously. Several have been tortured and killed for their efforts. The Committee to Protect Journalists reported that 53 journalists were killed and 325 were jailed in 1989.

Investigative reporting has been a dimension of the newspaper industry almost from the beginning, but became more common as commercial newspapers achieved large circulations. Reporters seek information from state officials that increases the sales of their newspaper and confers prestige on reporters. Reciprocally, representatives attempt to manipulate the flow of information to promote their causes and to increase the likelihood of their reelection. The relationships are simultaneously mutually beneficial and conflictual.

Newspaper reporters often seek a scoop, i.e., an attempt to outdo one another in obtaining information about behind the scene activities. Envy of and resentment toward elites is common among the nonelite of all societies. Newspaper reporters, editors, and publishers cater to those sentiments by emphasizing scandal and corruption not so much to attack elites, but to sell newspapers. Many state authorities look with askance at investigative reporting and resent the power of the press.

In representative democracies, legitimacy tends to become equated with public opinion as given voice by the news media. Public opinion, not the passions of kings, moves state affairs in representative democracies, and public opinion fanned by newspapers has unseated kings and presidents.

Numerous special committees have been commissioned to assess state/press relationships. Nearly all such committees assert that a viable democracy is contingent on an informed citizenry, but none have resolved the conflicts inherent to the relationship.

Following World War II, legislature-centered state structures and freedom of the press were instituted in Japan and West Germany and newspaper people led an international movement that called for all governments to adopt the principle of a free press. They claimed that the free flow of information would prevent wars and enhance the welfare of all.

Commercial newspapers reflect the biases of the marketplace as surely as state-operated newspapers reflect the biases of the state elite. The objective of commercial publishers, like that of all entrepreneurs, is to make a profit. If they attract readers they survive; if not they fail. If they can destroy competition they can gouge the buyers.

In the 1980s, approximately 60 percent of the space in U.S. newspapers was taken up by advertisements. Newspaper publishers cater to the interests of their advertisers as well as to the interests of their readers; sometimes these two sets of interests are in conflict, but not always: Many purchase commercial newspapers to learn which store is having a sale, not to acquire information about affairs of state.

NEWS IN TOTALITARIAN STATES

A few newspapers were established as organs of the states shortly after they appeared. But it was not until the twentieth century that states mounted major campaigns to transform nations via newspapers. Following the communist revolution, the USSR "imposed upon the printed word a uniformity more comprehensive and thoroughgoing than anything ever experienced in a more or less literate society" (Steinberg 1959:29). The Soviets not so modestly entitled their state newspaper *Pravda* (truth). In 1924, Mussolini placed all Italian newspapers under the control of the state and stated, "I consider Fascist journalism as my orchestra" (Ingelbart 1987:278). In 1933 all journalists in Germany became accountable to the minister of propaganda, who noted, "News policy is a weapon of war. Its purpose is to wage war and not to give out information" (p. 281). Some totalitarian states tolerate commercial newspapers but maintain close surveillance of them.

Whenever newspapers have been controlled by the state, news about how state policies are formulated, foibles of the state elite, and criticism of the current state structure have only limited circulation. Reports of impending meetings of state officials, of the deliberations of state officials, and of disagreements among state officials are uncommon in state-operated newspapers. One consequence is that whereas much of the information offered by commercial newspapers describes recent events, state-operated newspapers offer enduring truths and authoritarian dictums.

The continuation of totalitarian states without turmoil is impossible when news about state affairs is commodified. The elite of totalitarian states have had a chronic fear of the commodification of news from the beginning of newspapers. That fear continues unabated among modern dictators.

Commercial newspapers pander to the economic elite and the mercenary interests of readers; state newspapers pander to the state elite and attempt to educate readers. Commercial newspapers are frivolous; state ones are sober.

When newspapers are state operated, many suspect their reliability, but so long as citizens do not have access to other sources of information the definitions of reality offered by state newspapers are operational. In the USSR, for example, even during the 1950s more citizens regarded word-of-mouth information about state affairs as more reliable than that offered by newspapers (Inkeles and Bauer 1961), but there was no sustained movement to reform newspapers until the 1980s.

Nonetheless, even in totalitarian states newspapers increase awareness of state affairs by disseminating some information about the activities of state officials. For example, the reporters and editors of state-operated newspapers take as much pride in uncovering wrongdoings of state bureaucrats as do those of commercial newspapers. But comparisons and assessments of alternative state structures are rarely printed in state-operated newspapers.

NEWSPAPERS IN THE ELECTRONIC AGE

Newspaper circulation declined as the broadcast media matured. In 1967, 60 percent of the population of those between eighteen and twenty-nine years old in the United States claimed they read a daily newspaper. By 1988 that percentage had dropped to less than 30. The readership of Pravda markedly declined in the 1980s as a national television network was established in the USSR.

The maturation of broadcast news has enticed many newspapers to modify how they report the news. Some contemporary newspapers format news much as it is formatted for broadcasting. *USA Today*, established in 1982, offers news in a format similar to broadcast news. Lengthy accounts have been replaced by short items and state affairs are given less coverage; entertainment is given more.

The greater immediacy of broadcast news makes the news of newspapers "ancient history." In addition, the spoken word and dynamic visuals of television newscasts offer greater drama than the printed word and photographs. Perhaps within a few decades newspapers will become obsolete.

CONCLUSIONS

Prior to newspapers, temple and state elites defined communal reality. Commercial newspapers often offer definitions of reality that clash with those promulgated by temple and state elites. Unofficial definitions of reality compete with official definitions when the distribution of news is a commercial enterprise. Newspaper people were the first to seriously challenge the prerogative of state elite to define and interpret state affairs. As that occurred, the type of information that had currency changed. Among other things, information about the present supplanted information about the past. Commercial newspaper editors, reporters, and readers are not interested in information that will be judged significant in posterity; they are interested in exciting current events.

Many economic and literate elites as well as state and temple elites have opposed the dissemination of information about state affairs to the unwashed on the grounds that it leads to turmoil. Commercial newspapers have contributed to social turmoil. They have consigned state structures centered on a monarch or dictator to the dump heap of history. Representative democracy arose as an alternative.

Representative democracy rests on the belief that all have the prerogative to make assessments of state officials without fear of reprisal. That belief only has wide acceptance in societies with high rates of literacy and commercial newspapers. Competing assessments of the actions of state officials have had currency since the establishment of state structures, but the number of citizens who offered assessments of state officials remained very small until the development of commercial newspapers.

Most accounts of the transformations of state structures from palace-centered to legislature-centered ones posit the motive power as stemming from the downtrodden. That the downtrodden have on occasion rebelled is well documented. More often than not such rebellions have merely instituted another monarch or tyrant. Revolutions that have culminated in legislature-centered state structures supplanting palace-centered ones have, without exception, been led and organized by literates. It was journalists and pamphleteers more than any others who gave currency to the shortcomings of palace elites, awareness of alternative forms of state structures, and the dictum of the rights of man. Publics and competing political parties emerged only in nations with relatively unfettered commercial newspapers.

Representative democracy does not require that all be informed about state affairs, only that a substantial minority who are not fearful of reprisals be informed and have access to a mass medium.

Newspapers contribute to national unity in both democratic and totalitarian nations. In representative democracies, national unity is maintained through a series of negotiations that flow from confrontations between representatives advocating divergent interests and the distribution of information about those negotiations and their resolutions to citizens via news media. In modern dictatorships, national unity is reaffirmed by state-orchestrated ceremonies and news media are used to disseminate official definitions of reality.

Commercial newspapers are a form of social control as surely as are state-operated newspapers. They increase the accountability of state authorities to citizens and thereby facilitate control of authorities by citizens. Newspapers exercise control in much the same way as does gossip. Some deviants, both state officials and others, are mortified when their name appears in print. Gossip, whether spoken or in print, shames some of its targets into abiding by community standards. Newspapers serve as a collective conscience.

Printed information does not determine the thoughts of readers any more than spoken information determines the thoughts of listeners. The power of the news media stems from their ability to offer alternative definitions of reality to the multitudes. If the media offer a monolithic definition of reality, divergent thoughts are seldom formulated and even more rarely used to structure action.

10

Telecommunications: Overcoming Space

Telecommunication technologies, like newspapers, remove copresence as a necessary condition for communication, but unlike newspapers, they allow for bilateral relationships. Nearly everyone who occupies a central position in a modern social structure can quickly establish communicative contact with a multitude of distal persons. An entrepreneur in London can check his mailbox and answering machine for messages, fax a diagram to a branch office in Chicago, mail a postcard to his mother in Paris, telephone his wife in Stockholm, and then send a message by electronic mail to his business partner in Moscow. Chief executive officers of large organizations usually can establish bilateral communicative contact with most other members of their organizations in a few moments.

Some telecommunication systems are primarily unilateral, from a single source to a multitude. Publishers, broadcasters, and motion picture producers dispense information and entertainment to multitudes. Other telecommunication systems are bilateral from person to person, position to position, or some combination of person and position. It is the latter type of telecommunication that is analyzed in this chapter.

Orality, timekeeping procedures, numeric systems, and written languages are used to preserve information. They bridge time. Space-binding technologies, in contrast, are used to transmit information from location to location. They bridge space. With the exception of postal systems, telecommunications systems have for the most part been a twentieth-century development. The telegraph and telephone emerged during the nineteenth century, but telecommunication between private citizens via electronic media did not become commonplace until the twentieth century.

Prior to the development of the telegraph, the duration of the delay between the time when information was dispatched and awareness that the information had been received by the intended party was, with few exceptions, contingent on the distance that separated the communicating parties. Contemporary telecommunication technologies, however, make it possible for persons separated by thousands of miles to simultaneously dispense and receive information from one another.

Scholars and bureaucrats use telecommunication networks to share information; entrepreneurs and administrators use them to coordinate action; and, friends and families use telecommunications to affirm their relationships. Members of organizations as varied as the United Nations, Greenpeace, families, and employees of computer companies routinely share information and emotions, coordinate actions, and affirm relationships via telecommunication networks.

POSTAL SYSTEMS

As nations that encompassed large territories emerged, procedures for acquiring information from and transmitting it to the hinterlands were created. The specialty of "courier" appears in a list of professions compiled in the middle of the third millennium B.C. in Mesopotamia (Finkelstein 1979:52). In ancient Egypt, state functionaries known as the ears and eyes of the Pharaoh accumulated information about local conditions and transmitted it to the national capital; reciprocally the officials at the national capital dispensed directives to local officials.

The most ancient postal systems may not have used written languages. The Inca Empire included nonliterate specialists who raced on foot from way station to way station to transport spoken information to and from the national capital. As in the Roman Empire, the information transmitted in the Inca Empire via the postal system flowed from and to the national capital. The capital of the Chinese Empire also was the hub of a postal system. Whereas the Incas transmitted information via speech, the Romans and Chinese transmitted it in writing.

The ancient Chinese postal system, like other ancient systems, stressed secrecy (Cheng 1970:9). Illiterates usually served as couriers and the messages were sealed to assure that none but the sender and receiver were privy to the information. Couriers who broke the seals faced death. Nonetheless some couriers and supervisors "used their position to line their pockets" (p. 30).

The traders of some ancient civilizations may have established nascent postal systems to transmit information, but robust commercial systems that transported mail to private citizens first appeared in Europe during the Middle Ages. Governmental-operated postal systems that served the populace did not become common in any nation until the latter half of the nineteenth century. The relationships between the early commercial postal systems and the state elite of Europe were ambivalent, fluid, and often antagonistic. One British official of the sixteenth century noted, "With regard to correspondence conveyed by other messengers than their own, our Monarchs viewed it with great suspicion" (p. 7).

Nations with legislature-centered state structures were the first to establish state-operated postal services that promoted telecommunication among private citizens. The U.S. government established a postal network that included seventy-five postal offices late in the eighteenth century. In 1817, Senator Calhoun advocated the expansion of the U.S. postal service with, "Let us conquer space. It is thus that a citizen of the West will read the news of Boston still moist from the press" (as quoted in de Sola Pool 1983:77).

Newspaper publishers began using postal systems to transport newspapers almost as soon as post offices were established. In 1784, each day ten thousand London newspapers were delivered to other cities by the British postal service; ten years later the number had doubled. In North America, Franklin, the colonial postmaster and the leading newspaper publisher, took advantage of his position as postmaster to dispatch his newspapers to other cities. The use of postal services by newspaper publishers became so heavy that in 1843 the U.S. postal service banned the use of the postal service to distribute newspaper supplements.

The difficulty of extracting payment for services inhibited the growth of postal systems that transported mail from and to private citizens. The common procedure was for the recipients of mail to pay the postage, but many refused to pay. That difficulty was overcome when the adhesive stamp was invented in 1840. Prepaid postage was instituted in Britain the same year. Several other Western nations followed suit within a few years, and the amount of mail transported by post offices rapidly increased. In England, nineteen letters per person were handled by the post office in 1854; the ratio doubled by 1863. In 1896, the U.S. post office became the first to serve rural areas.

Whether or not the state should operate a postal system serving private citizens became a public issue in many other nations late in the nineteenth century. Agitation for a state-operated postal system that served private citizens arose among commercial interests in China in 1881 (Cheng 1970:95). Commercial newspaper publishers enthusiastically supported the development (p. 101), but many of the state elite opposed it. A postal system that transported private mail was initiated in 1886, and one thousand post offices were in operation in China by 1903 (p. 104).

The belief that citizens have the right to communicate with one another free from surveillance by the state authorities emerged as citizens began using postal services. Opposition to the opening of the mail of private citizens by state officials was fueled by newspaper editorials in many nations. Laws were enacted in the nineteenth century by the legislatures of United States, Canada, and France that prohibited state authorities from opening the letters of private citizens. Not all state authorities abided by the law, but the belief in the sanctity of correspondence between private citizens was widespread in Western nations early in the twentieth century.

The government is the heaviest user of the postal system in all nations. State officials use postal systems to disseminate information to citizens on everything from how to combat plant diseases to when taxes are due. With few exceptions, until well into the twentieth century the bulk of the communication between state authorities and citizens via postal systems was from state officials to citizens. The emergence of legislature-centered state structures has been accompanied by a partial reversal of the flow of mail between state officials and citizens. In representative democracies, citizens write to as well as receive mail from state officials.

All large commercial organizations use the mail to coordinate actions, advertise their goods, receive orders, and dispense products to customers. Mail order businesses were created in many nations shortly after postal systems that served the general public were established. As late as 1982, mail order businesses in the U.S accounted for 4 percent of retail sales.

Until recently, nearly all large postal systems were adjuncts of the state. In the past few decades several large commercial postal systems have been instituted. Most commercial postal systems specialize in transporting information and small objects for other commercial organizations. Commercial postal systems have eroded the monopolies of state-operated postal systems. In addition, electronic telecommunication systems now compete with postal systems for customers. Although postal systems no longer are the dominant form of telecommunication, nearly all citizens of wealthier nations check their mail boxes every day, and most office workers open their office mail shortly after arriving at work.

THE TELEGRAPH

The telegraph divorced the transmission of information from the transportation of objects. Prior to the invention of the telegraph, with the exception of a few devices such as optic telegraphs, the transmission of information had been constrained by the transportation of objects.

Until the nineteenth century, the rapidity of transmitting information across space had remained much the same for millennia. The pony express transported letters somewhat more rapidly than the postal system of the Roman Empire, but not much. Until the invention of the railroad, with few exceptions, the rapidity of the delivery of information to distant locations was limited to that of a speeding horse.

Drums, smoke signals, and flashing mirrors, and a few optic networks had been used to transmit information across space on a limited scale prior to the invention of the telegraph. France instituted a nationwide optic network in 1793. The developer of the system wrote, "The day will come when the government will be able to achieve the greatest idea we can possible have of power, by using the telegraph system to spread directly, every day, every hour, and simultaneously, its influence over the whole republic" (quoted in Attali and Stourdze 1977:97). An optic network linked New York and Philadelphia in the nineteenth century. The weather often rendered these systems inoperative, however, and they became obsolete almost as soon as they were perfected.

The U.S. Post Office began installing telegraph lines in 1844 and approximately thirty thousand miles of telegraph lines were put into service within a year. The network was sold to a commercial company in 1846. Two decades later Western Union emerged as the first gigantic commercial communications monopoly.

The perfection of the transportation of goods by rail and the transmission of information by the telegraph coevolved for a few years in the United States. The telegraph was used to transmit information about the arrival times of trains from station to station. Railroads were a quantum leap in transporting goods; the telegraph a quantum leap in transmitting information.

Reporters, traders, and state officials began using the telegraph almost as soon as the networks became operational. Newspaper reporters dispatched stories to their editors, and the editors transmitted news from city to city and sent instructions to reporters via the telegraph. Prior to the laying of a trans Atlantic cable in 1868, news about European affairs published in North American newspapers was a month or two old. After the Earth was blanketed with telegraph networks, information about events that occurred prior to yesterday was no longer news. As the telegraph networks expanded, international news became a larger segment of newspapers, so reflecting Samuel Morse's earlier assertion that the telegraph would "diffuse, with the speed of thought, a knowledge of all that is occurring throughout the land; making in fact, one neighborhood of the whole country" (quoted in Czitrom 1982:12). Sad to say, the telegraph did not usher in Utopia.

In the United States the telegraph was conceptualized as a commodity subject to the same restrictions as other commodities. In the early decades, several controversies over use of the telegraph got to court, but none was argued in terms of the First Amendment. In contrast, the state authorities of most nations conceptualized telegraphic transmission as a form of communication and assumed control of telegraphic networks. For example, commercial telegraph companies were established in Britain shortly after the telegraph was invented, but in 1869 Parliament authorized the post office to purchase the telegraph network and establish a nationwide system.

In most nations, the telegraph networks were of greater benefit to state officials than to ordinary citizens. State authorities used them to more rapidly acquire information from and dispatch orders to the hinterlands and to more rapidly and effectively marshal coercive forces. Some state-operated telegraphic networks were available to citizens, but often the telegrams were subject to censorship. For example, in the early years, all telegrams in France had to be approved by a state censor before they were transmitted. Some private citizens, especially newspaper reporters and traders, used telegraph networks to transmit information, but telegraphic communication had relatively little impact on the frequency of communication among private citizens who lived some distance from one another. Prior to the development of telephonic networks, family members and friends who lived apart and maintained contact with one another continued to do so for the most part through the mail.

THE TELEPHONE

The dream of communicating across space via speech was transformed into an actuality in 1876. The specialists in telegraphic communication were not impressed by telephonic communication, however, and by and large regarded the telephone as an electronic toy. At that time, the United States was "crisscrossed by 214,000 miles of telegraph wire delivering 31,703,181 telegrams through 8,500 offices" and plans had been formulated to link private homes to the telegraph network (Aronson 1977:18). Nearly all involved in telegraphic communication presumed the telegraph would remain *the* procedure for rapidly reciprocating telecommunications.

Although Alexander Graham Bell advocated the telephone as an extension of interpersonal communication, it was not apparent to all that its greatest potential was bilateral telecommunications. In Hungary, a telephone system was installed as a unilateral broadcasting medium in 1893 to transmit news and music to subscribers, but its potential for bilateral communication was not widely shared.

A stumbling block to developing large-scale interpersonal telephonic communication was the logistics of connecting the outlets with one another. One wire allows communication between two locations, but it is necessary to have ten wires for station-to-station contact among five locations and twenty-one wires for station-to-station contact among seven locations. Manually operated centralized exchanges partially overcame that difficulty. The first centralized exchange was installed in 1877 (p. 24), with merchants as the first subscribers. The Pittsburgh telephone directory of 1879 lists three hundred telephones, all of which were places of business or the homes of businessmen (p. 27).

The state authorities of most nations assumed control of telephone systems. A British court ruling in 1880 "pronounced a telephone conversation to be a telegram" and the government assumed operation of the telephone network. Britain's telephone system was routinely subject to scathing criticism in the newspapers in the following decades (Perry 1977:80–81). The French government assumed control of the telephone network on the grounds that "[g]overnments have always kept to themselves the exclusive use of things which, if fallen into bad hands, could threaten public and private safety" (Attali and Stourdze 1977:98). Some governments that operated telephonic networks promoted telephonic communication among private citizens; others were indifferent to it; and still others suppressed it.

The cost of telephonic communication has declined rapidly in the past few decades while postage has increased. For example, in the United States in the 1930s it cost three cents to mail a letter a thousand miles, while a brief telephone conversation would cost a few dollars. In the 1990s, one can speak for a minute on a long distance call almost as cheaply as one can mail a letter.

Until recently nearly all telephonic communication was conducted via stationary telephones. Cellular telephones now allow people to make telephone calls from almost anywhere as they travel from location to location. Access to telephonic networks is approaching universality, as on a worldwide basis there exists almost one telephone for each adult.

TELEMATICS

The term *telematics* is a derivative of the French word *telematique*, which in turn is a compound of *telecommunication* and *informatique*. The first step toward telematics was taken when printing machines were attached to telegraph receiving units in 1859 to make printouts. Facsimiles of configurations were transmitted in 1865. The linking of electronic telecommunication procedures to computer-processed information in the 1960s ushered in modern telematic procedures. Information about the stock markets was first transmitted via telematic procedures to newspapers in 1962. The airline industry was the first to routinely transmit information via telematics. Until the 1970s, nearly all telematic transmissions were domestic, but such transmissions across national boundaries have grown by leaps and bounds in the past decade.

The merger of telecommunications with computers and recorders rendered the distinctions between transmitting information, information processing, recordings, broadcasting, and print problematic. One U.S. court case revolved around "whether a newspaper delivered by teletext is an extension of print and thus as free as any other newspaper, or whether it is a broadcast and thus under control of the government" (de Sola Pool 1983:2). As these distinctions have become blurred, computer companies have began competing with telecommunication companies for customers.

Telematic procedures transmit information far more rapidly and at far less cost than traditional telecommunication ones. For example, the contents of a large book can be transmitted halfway around the world very rapidly at very little cost (Blatherwich 1987:4). Telematic procedures have also extended how communication between dispersed persons can occur. For example, the merger of telephonic networks with recording procedures has produced the hybrid known as voice mail. Answering machines, like mail, preserve information as well as facilitate the transmission of information across space.

Telematics are such an integral dimension of most complex social structures that the well-being of nations is contingent on their effective use of telematics. In 1979, a Canadian commission noted, "If we wish to control our economy then we will require a sophisticated telecommunications" (p. 25). In 1985 another Canadian commission called on the government to alert citizens of the perilous position of Canadian sovereignty (p. 26) due to intrusion of foreign telematic systems into Canada.

The transformations associated with the development of telematics may prove to be as sweeping as those associated with the emergence of commercial newspapers. Just as the printing of news required only minor modifications of the procedures used to print books, the merger of computers with electronic telecommunication networks required only minor modifications. And, just as the development of commercial newspapers greatly increased the amount of information about state affairs in circulation, telematics greatly facilitate the dissemination of vast quantities of information. Telematic machines linked to international networks may become household items in wealthy nations within a decade.

ENHANCING SOCIAL STRUCTURES

All contemporary chief executives of large organizations are at the hub of a complex telecommunication network. When it is impossible to maintain a bilateral flow of information within a large structure, the structure collapses. If all the messengers of a preelectronic nation were incapacitated, the nation would collapse. Likewise, if all the electronic telecommunication machines of a modern nation malfunctioned, the nation could not function.

Telephones, two-way radios, intercom systems, and telematics have largely supplanted messengers as the primary means of transmitting information within and between complex social structures, and they have allowed for a tremendous growth in size and number of complex organizations. A few multinational organizations spanned the world prior to the development of electronic telecommunications, but compared to modern multinational structures they were rather modest endeavors.

Electronic telecommunication networks allow the chief executive of large social structures to acquire information from and issue directives to all segments of an organization almost instantaneously. The president of the United States can acquire information from almost all other members of the state structure in minutes; and, reciprocally he can alert nearly all citizens to an impending catastrophe within minutes.

Electronic telecommunication systems allow chief executives to monitor on-line personnel and issue directives to them without relying on intermediaries. For example, the chief of a large metropolitan police force can use telecommunications to acquire information about what is transpiring in a distant part of the city and to assume command of the on-site personnel. Electronic telecommunication systems allow chief executives to collapse multilayered organizations into two-level structures with one level consisting of the chief executive and the other layer consisting of all other members of the organization. Local authorities of complex organizations accordingly have less autonomy than those of a few decades ago.

Ancient monarchs, such as King David of the Old Testament, occasionally exerted their wills on distal subordinates through written orders, but there always was some delay between when the orders were dispatched and when they were received, and usually there were intermediaries who issued the orders to the local on-line personnel. In contrast, modern executives can use telecommunications to issue orders directly to on-line personnel far removed from headquarters. For example, Hitler used the telephone to issue orders directly to field commanders. When the German generals organized a rebellion against Hitler in July 1944, they placed many of the Nazi elite under house arrest but failed to cut their telephone lines. The Nazi leaders used the telephone network to coordinate the actions of loyal units and crushed the rebellion within a few hours.

If chief executives are to have effective command of distal subordinates via telecommunications, the network must be contextualized by a social structure that makes distal subordinates accountable to their chief executive. For example, toward the end of World War II, Hitler issued an order to the general in charge of the German army defending Paris to burn the city when the German army fled. As the Allied forces approached Paris, Hitler repeatedly telephoned the general to inquire, "Is Paris burning?" The general disobeyed Hitler's orders.

Although the intent to enhance extant structures informed the elaboration of nearly all telecommunication technologies, their development sometimes eroded boundaries and hierarchies by making it possible for subordinates to communicate to their superordinates and to one another without becoming known to them. Anonymous communication allows the lowly to intrude on the mighty without fear of reprisal. Nearly all complex organizations contain arrangements that free chief executive officers and other high-ranking authorities from the possibility of being the recipient of anonymous contact via telecommunications. Presidents, generals, and other chief executives usually are protected by receptionists who screen their letters and calls, determine the identity of those attempting to establish contact, and perhaps the topic to be addressed before authorities become available to those attempting to communicate with them via telecommunications.

On occasion, telecommunication systems have been elaborated to empower rank-and-file citizens. Several governments extended their postal system to the general populace to empower citizens, and some governments have justified taking control of telephonic networks on the grounds that government-operated systems will be more responsive to the needs of citizens than would commercial systems. The expansion of telecommunications has empowered citizens to be more effective in their calls for services and rendered officials somewhat more accountable to citizens. In the large cities in the United States in the 1980s, the number of telephone calls received by police stations each year was equivalent to 75 percent of the Britain the comparable figure was 25 percent (Manning 1988:7). When a significant public opinion issue arises in legislature-centered states, representatives are deluged by telephone calls, telegrams, and more recently electronic mail.

FORMING SOCIAL STRUCTURES

Telecommunication technologies have extended the ability of human beings to associate with one another on the basis of affinity and rendered proximity less significant. A substantial percentage of the world's population has associations with others whom it has never met. Voluntary associations ranging from grammar school pen pals to the International

Sociological Association are maintained via telecommunications. Members of such organizations may seldom or even never establish copresence.

Each voluntary association is of very limited significance, but in their totality they constitute a significant dimension of the social structure of the modern world. Voluntary associations populated by dispersed persons who share an interest and communicate with one another via telecommunications have the potential of impacting on other social structures. Many voluntary associations began with spatially separated people who have a common interest using telecommunications to establish contact with one another, strengthen their bonds with one another, develop distinctive definitions of reality, and create a collective identity. Once those elements of sociation are in place, the organization then can respond to and act toward other social units.

Most monarchs and dictators have recognized that voluntary associations are a threat to asymmetric state structures and that telecommunication technologies enhance the likelihood of voluntary associations. Consequently many of them prohibit autonomous voluntary associations and most have restricted the availability of telecommunication technologies to rank-and-file citizens. When Stalin assessed the utility of the telephone he stated, "I can image no greater instrument of counter-revolution in our time" (as quoted in Dizard and Swensrud 1987:1). The leaders of the 1991 revolution that overthrew the military dictatorship of Thailand were known as the mobile phone mob.

The paranoia of dictatorships toward the extension of telecommunication procedures to the populace is well placed. When the subordinates of an organization have the opportunity to communicate with one another without convening, it enhances the likelihood that they will become a threat to extant hierarchies. Such has been the case from at least the time of the American Revolution, when the revolutionists used the colonial postal system to distribute subversive ideas, up to the 1989 uprising in China, when students used telematic procedures to issue rallying cries. It is impossible to maintain a dictatorship when all citizens have access to telecommunication networks.

The vast majority of voluntary associations do not become active in political affairs. But even those which do not contribute to the formation of alternative definitions of reality often compete with definitions of reality offered by traditional elites. For example, the first voluntary associations that rested on a foundation of telecommunications were the scholarly associations that emerged in Europe in the sixteenth century. The printing press allowed those scholars to publish their letters and the emerging postal systems provided a procedure for them to transmit their letters to one another. The definitions of reality offered by those scholars competed with and in the long run overcame traditional definitions of reality.

FORMS OF SOCIATION

Telecommunications technologies allow separated persons to create a shared virtual reality and to create a cybernetic presence. Even the most sophisticated telecommunication technologies do not allow people to establish tactile contact, or see or hear one another. Telecommunicators can only see objects and events that replicate other objects and events or hear sounds that attempt to replicate other sounds. When we speak to a friend via the telephone we sometimes assert, "It's nice to hear your voice." But we do not hear our friend's voice, only a replication of it.

To a large extent the perfection of telecommunications has consisted of movement from low fidelity of the shared virtual realities created by telecommunications to high fidelity. The experiences one has when one speaks to an acquaintance via the telephone has far greater similarities to the experiences that are obtained when the other is copresent and speaking than are the experiences derived from a letter.

Video telephones have further enhanced the fidelity of shared virtual realities created via telecommunication. Video telephones make it possible for telecommunicators to establish shared visual foci and to simultaneously use discourse to inform one another about the shared foci. For example, two architects separated by thousands of miles can analyze a diagram displayed on terminals in front of each of them and make modifications in the diagram on the basis of simultaneously being aware of the modifications each of them makes.

The ability to establish a cybernetic presence via telecommunication is so taken for granted by contemporaries that most are no longer impressed by it. However, when each technique for creating a cybernetic presence was invented, some regarded it as a savior of humanity. Another indication of the distinctiveness of a cybernetic presence is suggested by the reactions of adults who were unaware of such a possibility upon learning that a cybernetic presence can be created.

When adults who were unaware of writing first become aware of the ability of people separated by space to communicate with one another via "talking leaves" they are awed by the accomplishment (Harbameier 1988:255). One account relates that a European wrote a note on a wood chip and asked a nonliterate to deliver the message to the writer's wife. The nonliterate asked the author what he was to say to the wife and was told that it was unnecessary to say anything. The nonliterate asked, "How can this speak? Has it a mouth?" The message was delivered and in the following days the messenger was often seen "surrounded by a crowd who were listening with intense interest while he narrated the wonders which this chip had performed" (Clammer, as quoted in Scribner and Cole 1981:3). When the Cherokees became literate, young men "went on long journeys for the sole purpose of writing and sending back letters to their friends" (Foster as quoted in Harbameier 1989:266).

Prior to the invention of the telephone, disembodied speech was regarded as a supernatural phenomenon. Devils, the dead, and God on occasion were reported to have talked with people. Communication via disembodied speech has become part of the taken-for-granted for nearly everyone.

Prior to the emergence of telecommunication, the construction of shared pasts and shared futures was restricted to those who established copresence. Now, of course, those who can establish a cybernetic presence can create shared pasts and shared futures. A multitude of people have established contact with one another via one or more telecommunication technologies, communicated about a number of topics, and projected a shared future of convening. When they convened, their encounter was contextualized by the shared pasts and shared futures that they constructed via telecommunications.

Each type of telecommunication network favors and disfavors the construction of particular forms of thought and action. For example, letters, telegrams, and E-mail separate the composition of messages from their transmission and favor the construction of carefully formatted messages. Other procedures such as telephonic communication favor rapid-fire bilateral responsiveness and disfavor the construction of carefully formatted messages. Consequently it is more difficult to negotiate via the mail or telematics than via the telephone. Most who make extensive use of telecommunications prefer telephonic communication when the intent is to negotiate agreements and telematic transactions when the intent is to transmit instrumental information from one location to another. Telematic procedures deliver "hard copy"; telephone conversations offer "soft copy."

CONCLUSIONS

The obliteration of space has been the implicit objective of the developers of telecommunication technologies since the first postal system was put in place. Space has not been obliterated, but space is no longer the barrier to the sharing of information and sentiment that it once was.

Telecommunications emerged from and was used to maintain national

structures for millennia but began to become unshackled from state structures in the nineteenth century and some forms acquired a high degree of
autonomy in the twentieth century. Some state authorities continue their
attempts to control telecommunications but find it almost impossible
to do so.

Telecommunication technologies are eroding nationalism. Two sisters,
one in Saigon and the other in Sidney, may regularly converse with little
or no consciousness of their nationality. A telephone conversation between two persons typically is of little import to anyone other than the
two conversationalists. But when millions of people acquire information,
affirm relationships, and program shared futures via telecommunication
networks the significance of space-bonded relationships, such as nationalism, recedes.

The structures of the commercial telecommunications networks probably are at least as significant a part of the social structures of the world as
are traditional state structures. One's location on international communication networks is becoming more significant than one's geographic location. Access to telecommunication networks has become so widespread
that it is extremely unlikely that they can be brought under the control of
state structures.

Telecommunication networks have rendered proximity almost irrelevant for the formation of sustained associations. Pen pals have partially
replaced playmates for many grammar school students; among scholars,
E-mail correspondents are replacing departmental colleagues.

Although telecommunications have contributed to social unity for millennia, only recently have they made more than minimal contributions to
cultural homogeneity. The postal system of ancient Rome was the linchpin
of the Roman state structure but did little to promote cultural homogeneity.
Currently vast quantities of information and entertainment are transmitted
around the world and nearly all, especially the younger generation, expose
themselves to information and entertainment from afar.

Morse's telegraph was the first toddling step into the age of electronic
communication; Babbage's analytical machine the first toddling step into
the computer age. For almost a century the two technologies matured
independently. They are now unified. The transformations wrought in
social structures by their unification may equal, if not surpass, those
associated with the maturation of newspapers.

11

Recorded Sounds and Sights

Although printed books have been available for five and a half centuries and recordings for only little more than a century, more people listen to and view recordings than read newspapers, magazines, and books. Compact disks, CD players, cassette tapes, and cassette players can be purchased in nearly all communities. Most schools, governmental agencies, and modest-sized commercial companies have a library of recordings. Most of the sounds and sights broadcast by radio and television stations are prerecorded.

Each day millions of families memorialize celebrations by recording them, tens of thousands of commercial companies use recorders to maintain surveillance of their retail and wholesale outlets, thousands of reporters record newsworthy events to obtain sound and sight bites for newscasts, and hundreds of researchers use audiovisual recording devices to generate and preserve data. Students from grammar school through graduate school experiment with recording sounds and sights. Documentaries have partially replaced documents as repositories of knowledge.

The experiences we derive from information technologies have become more dramatic and fluid as people have turned from written languages and print to recordings to preserve and dispense information and entertainment. Written languages and print favor static referential symbols; recordings favor dynamic evocative symbols.

THEIR DEVELOPMENT

The immediate impetus for recording dynamic sights was a dispute over whether all four feet of a galloping horse were off the ground at the same time. The argument was settled in 1877 when Muybridge, a professional photographer, took twenty-four photographs per second of a galloping horse. During the preceding decades, still photography had achieved widespread popularity. A technique for recording dynamic sights on celluloid was developed within a year. Four years later the kinescope (the original motion picture) was patented in the United States and amusement parlors began to install them.

Machines that transformed sound waves into impressions on cylinders of wax and those impressions back into sounds were invented in 1878. Audio recording devices also quickly achieved popularity in amusement parlors and became household items in many nations within a few decades.

The developers of the early recorders and playback devices attempted to create machines that replicated sounds and sights as precisely as possible. Nearly everyone marveled at the verisimilitude of the sounds and sights offered, although by contemporary standards the fidelity of the early recorders was low.

Adventurers and entrepreneurs traveled far and wide to record exotic sights. Screenings of early motion pictures were usually accompanied by a speaker who explained each scene as it appeared on the screen (Barnouw 1974:21). Documentary motion pictures were a popular source of entertainment in the latter part of the nineteenth century in Europe, the Americas, and several Asian nations. They "outnumbered fiction films as late as 1907" (ibid.).

The production of fictional motion pictures began to flourish in the 1910s. Hollywood emerged as the center of the motion picture industry after World War I and the construction of motion picture theaters was a booming industry in North America and Western Europe throughout the 1920s. Inventors and entrepreneurs formed coalitions to merge audio and visual recording techniques and success was achieved in the early 1920s. Motion pictures that included recordings of the sounds of thunder storms and stampeding cattle as well as speech quickly became standard fare at motion picture theaters. In 1927 there were twenty motion picture theaters with sound in the United States; by 1931 there were over thirteen thousand. In both France and Germany the last silent feature length motion pictures were produced in 1929 (Monaco 1976:12). Attending motion picture theaters became the most popular form of entertainment in North America, Western Europe, and parts of Asia in the 1930s (Monaco 1976). In 1938 eighty million Americans, equivalent to 65 percent of the population, attended motion picture theaters each week and the motion picture industry was the fifth largest industry.

The glory years of motion picture theaters was short-lived. The development of television broadcasting eroded the appeal of motion picture theaters. For example, in the United States by 1973 only twenty million, equivalent to only 10 percent of the population, attended motion picture theaters. The decline in attendance at motion picture theaters was somewhat slower in most other nations, but as first television broadcasting became more widespread followed by the perfection of magnetic cassette recordings the appeal of motion pictures declined around the world.

Radio stations began to use magnetic devices in 1947 to record sounds and television stations began to use magnetic recorders to preserve dynamic sights in 1956. The practice of using magnetic recordings spread slowly for a few years but within a decade many radio and television stations began broadcasting magnetically recorded programs. Several commercial corporations began using magnetic audiovisual recordings in the 1960s and in the 1970s producers of pornographic motion pictures began to use magnetic recordings instead of film (Klopfenstein 1989:30). In the 1980s magnetic recording and playback machines became household items in many nations.

The income from video sales surpassed the revenues from motion picture theaters in both Spain and Sweden in 1984 (Spa and Jones 1986:36). About 1 percent of the income of the Hollywood motion picture studios was derived from magnetic recordings in 1980, but by 1987 videocassette sales surpassed the income from that of motion pictures theaters in the United States (Marlow and Secunda 1991:132) and the production and distribution of video recorders, playback units, and cassettes became big business worldwide.

Japan exported over 22 million video recorders in 1984 (Usami 1988:72). By 1989 in the United States, the recording industry was second only to the aerospace industry in the amount of money generated by exports by 1989. The British took the early lead in the production of pirated videocassettes, but by 1990 Britain was challenged by several other centers for leadership in the production of pirated cassettes. Videocassettes of feature length motion pictures are available at a modest cost worldwide (Alvarado 1988). Magnetic recordings are less expensive than film to produce, can be played back immediately, and little training is required to use them. A substantial percentage of the cassettes are pirated productions and a substantial percentage of recorders, playback units, and cassettes are distributed via the black market.

RELIGION AND STATE CONTROL

Religious and state authorities have paid little attention to audio recordings, but as the motion picture industry matured, some attempted to control their distribution. For example, the Catholic church published lists of disapproved motion pictures and a few religious groups banned the viewing of motion pictures. However, the lure of motion pictures has been so great that most religious prohibitions have been ineffective. State authorities, often acting at the behest of religious interests, were relatively successful in controlling the distribution of motion pictures preserved on film, but the perfection of magnetic recording devices has severely eroded their ability to control the distribution of motion pictures.

Many state authorities did not react to the magnetic recordings of motion pictures until after a substantial percentage of citizens had acquired playback units and a sizable number of videocassettes. In most regions of the world, those with the financial resources can purchase a videocassette of almost any motion picture that has ever been marketed.

The authorities of several nations attempted to harness motion pictures to their own interests. The Nazis were especially active in using motion pictures to promote their definitions of reality. Ironically, German motion pictures were submitted as evidence against Nazi officials at the Nuremberg trials, and motion pictures of the German concentration camps were widely circulated to document the horrors of Nazism. In the 1920s, the USSR placed the production of motion pictures under the control of a state agency charged with the task of promoting the interests of the state. In the 1980s the state authorities of China undertook large-scale production of videocassette motion pictures to promote their interests (Ganley and Ganley 1987:84).

The relationship between the state authorities of legislature-centered states and the distributors of motion pictures has been fraught with ambiguity and inconsistencies. Nearly all legislatures have debated the merits of recordings, especially motion pictures, and on occasion have banned those judged a threat to the public welfare. The efforts of the authorities of legislature-centered states to suppress the circulation of recordings have been even less successful than the efforts of dictatorships.

In a few nations motion pictures were conceptualized as an economic enterprise subject to only the same controls as those applied to other economic endeavors. For example, in 1915 the U.S. Supreme Court declared that the motion picture industry was a business "pure and simple" and not protected by the First Amendment (Jowett and Linton 1980:71). In 1922 the Hollywood producers formed an association to protect themselves from state controls and formulated a production code. Several domestic entrepreneurs and importers of films violated the code. In the 1950s the U.S. Supreme Court began to apply the First Amendment to motion pictures and several restrictions were lifted. The relationship between the state and motion picture industry in many other nations with legislature-centered state structures evolved in a similar manner.

Many of the efforts of state authorities to control the distribution of motion pictures have boomeranged. For example, when the state authorities have undertaken the production of motion pictures, it has often been more effective in whetting the appetites of their citizens for motion pictures other than those produced by the government than it has been in generating support for the state authorities and their policies. This was the case when the Soviet government produced thousands of motion pictures to hasten the arrival of Utopia and legitimate the communist dictatorship, but the effort did not contribute to either the arrival of Utopia or the stability of the Soviet dictatorship.

Thousands of petty entrepreneurs produce pirated replicas of recordings and distribute them via worldwide black markets. Much of the pirating is done to avoid paying royalties and taxes but some recordings are pirated to avoid censorship. The percentage of recordings that are pirated is difficult to estimate, but currently there probably are as many pirated videotapes in circulation as there are legitimate ones.

In some nations the authorities only make token efforts to control the black market distribution of pirated recordings. Even when the police raid illegal video establishments it has little impact on the number of pirated recordings that are distributed. Black markets of recordings flourish in nearly all nations where the state authorities attempt to control their distribution.

The international black market networks for distributing videocassettes were initiated by travelers, migrants, and privileged elites (Ganley and Ganley 1987:51). In the 1980s, the flights from Singapore to India became known as the VCR flights among the flight crews in recognition of the many passengers carrying recorders and cassette into India. One estimate is that in the mid 1980s at least 20 percent of videocassette recorders were smuggled into India to avoid paying taxes (Agrawal 1988:91). It is also estimated that about 90 percent of the cassette tapes sold in India were pirated (p. 97). In the late 1980s, five times more videocassette recorders entered the United Arab Emirates than there were residents of the nation. It is almost impossible to determine the ultimate destination of the recorders, but it is likely that most of these machines found their way into surrounding Muslim nations where they were banned (Ganley and Ganley 1987:22).

When state authorities attempt to control the distribution of cassette tapes, it is a common practice for those who own VCRs to share tapes, and some owners charge their neighbors a modest fee to view recordings in their homes. In regions where many are too poor to own a VCR, coffeehouse and tavern owners rent tapes of motion pictures to show their customers. Once a magnetic recording has penetrated a national boundary, it is almost impossible to prevent its replication and distribution. It is as if the development of the printing press had been accompanied by the invention of copying machines.

One indication of the difficulty that state authorities have in controlling the distributions of cassette tapes is that in 1980 the Saudi authorities attempted to prevent the BBC from broadcasting the motion picture *Death of a Princess* and banned it from Saudi Arabia. A recording of the broadcast was flown to Saudi Arabia the next morning, duplicated, and distributed via the black market in less than twenty-four hours after it had been broadcast in Britain. Pirated replications of American motion pictures often reach Kuwait via Singapore before they are shown in American theaters (Boyd and Adwan 1988:169). Many recording artists and motion picture stars are as popular in nations where their recordings are banned as they are in nations where their recordings freely circulate.

Foreign video tapes and VCRs were banned in the USSR almost until its collapse, but cassettes of foreign motion pictures were widely distributed in Soviet cities throughout the 1980s. In 1987, the Soviet government began manufacturing videocassette recorders that were incompatible with Japanese and Western recordings. Petty entrepreneurs quickly emerged who specialized in making them compatible for a modest fee (Boyd 1989:215).

In several nations, state authorities and wealthy citizens reached a tacit agreement that the distribution of recordings would be tolerated if censorship of the broadcast and print media were tolerated. When the state controls the broadcast media many of the wealthier citizens turn to pirated videocassettes for entertainment and information. On occasion the prevalence of recordings has become so great that state officials have been enticed to give greater autonomy to the broadcast media. For example, the Malaysian government allowed a commercial television broadcasting station to be established to counteract the black market in recordings (Ogan 1989:52).

Several factors influence the pervasiveness of VCRs and videocassette recorders in nations. In many wealthy nations where state authorities control television and severely censor motion pictures, most homes have a VCR. The majority of the households in Saudi Arabia possessed a VCR in the mid-1980s; in preinvasion Kuwait 80 percent of the households owned at least one VCR (p. 49). In contrast, in 1984 only about 2 percent of Italian homes had a VCR (Thibault and Toriano 1988:50). Several factors contributed to the lack of popularity of VCRs in Italy: for example, during the time VCRs achieved popularity on a worldwide basis, the Italians already had a wide variety of television shows to choose from.

Some cassette recordings are distributed to stir political and religious sentiments. Prior to Ayatollah Khomeini's return to Iran, his supporters smuggled in audiocassettes of his speeches (Ganley and Ganley 1987:9). His followers and potential recruits to the cause gathered in homes and coffeehouses to listen to the recordings. When reformers challenged Ferdinand Marcos in the Philippines, "videotapes of foreign newscasts about the Aquino assassination and Marcos' corruption" were shown in homes and at political rallies (Straubhaar and Lin 1989:126). Political sentiments also have often been stirred by recordings intended primarily for entertainment. For example, two of the most popular motion pictures distributed via the black market in Iran in the 1980s were *Footloose* and *Return of the Jedi* (Ganley and Ganley 1987:67). *Footloose* offers the story of a youth's rebellion against a fundamentalist religious sect. Many Iranians equated the character Darth Vader in *Return of the Jedi* with Khomeini.

The authorities of dictatorships have been more obsessed with controlling the distribution of audiovisual recordings than with controlling the distribution of newspapers and books. Apparently they believe that subversive ideas can be more effectively transmitted via recordings than print. That belief may be well founded but there is little research to document it.

Many claim that the distribution of recordings from the electronically advanced to the less electronically advanced nations is a form of cultural imperialism that undermines domestic well-being. International trade in recordings enhances consciousness of alternatives and undermines nationalism, but that it undermines the welfare of the citizens of the importing nations has not been demonstrated.

Videocassettes frequently are the most common source of information about current affairs for illiterates in poor nations. In legislature-centered nations with high literacy rates, candidates for national offices devote more resources to the production videocassettes to convey their message to voters than any other campaign tactic. The broadcasting of video recordings has become the dominant campaign tactic in nearly all democratic nations.

ENTERTAINMENT

The most common use of recordings is to provide entertainment, and recording technologies have immensely increased the amount and variety of entertainment available. Motion pictures are a source of excitement for nearly all, from the elegant elite to the very poorest. When recordings first became available, nearly all were entertained by the marvel of machines replicating sounds and dynamic sights. The ability of recordings to provide replicas of high fidelity is now taken for granted by nearly all, although some continue to concern themselves with enhancing the fidelity and vividness of the replicas offered.

Film and video producers have developed many techniques to intensify the experiences of audiences. They use close-ups to lessen the conventional distance between the viewer and the viewed, which magnifies the emotions generated by evocative gestures displayed by recordings. Camera angles, editing, and juxtaposing render space and time elastic. The concept of "scene," taken from live theater, was replaced by the concept of "shot." Just as scenes are shorter and more focused than events from mundane life, shots are shorter and more focused than theatrical scenes. Recording and playback devices allow events to be speeded up, slowed down, frozen, or reversed. The producers of recordings have established their own canons of production and offer fantastic sights and sounds that sometimes elicit larger-than-life emotions. Events that flow from the imaginations of producers often are far more pleasing and vivid than those obtained from the mundane world.

Motion pictures provide new opportunities for people to suspend their disbelief and enter the world of fantasy. When listeners, readers, and viewers suspend their disbelief, they vicariously share experiences with those who offer the narratives and reflexivity recedes. The extended suspension of disbelief is highly contingent on performers, writers and producers offering flowing and well-crafted products. Disjunctive productions, whether poems, novels, or motion pictures heighten the self-consciousness of audiences and inhibit them from suspending their disbelief. Poorly crafted motion pictures elicit giggles instead of thrills.

Although motion pictures have migrated from motion picture theaters to living rooms via videocassettes, millions of people continue to turn to them to enter the world of fantasy. Advertisements of next week's motion picture do not elicit the intense joyful anticipation that they did during the early decades of motion picture development, but Hollywood remains the world's leading dream factory.

The experiences of audiences of live performances are a combination of the relations between the performers and audiences, relations among members of the audience, and the content of the performance. In contrast, the experiences of persons listening to and viewing recordings in solitude are largely derivatives of the content of the recording. The interaction between performers and audiences and within audiences at live entertainment has been replaced by solitary absorption in the sounds and sights as people turn from live performances to recordings for entertainment.

DOCUMENTARIES

Most recordings are made to entertain, but a substantial percentage of them are made to inform. Recordings made to entertain are designed to elicit emotions. In contrast, recordings made to inform are designed to replicate sounds and sights. Although recordings offered as entertainment also inform and those made to inform often entertain, in general, esthetic criteria are applied to recordings designed to entertain whereas questions of authenticity are directed at recordings designed to inform.

The earliest recordings attempted to replicate events, and the enjoyment derived from them stemmed largely from the marvel of replication. The dichotomy between entertaining recordings and documentaries began to emerge as the producers of recordings learned to control recording procedures: they began to use editing techniques to elicit particular emotions from audiences and other procedures to preserve replicas of events.

In one sense, the first documentary was produced in 1874 when Venus's transit of the sun was recorded in France by a *revolver photographique*. A book advocating the use of motion pictures to document historical events was published in 1898 (Barnouw 1974:26–27). A 1903 film displayed the circulation of blood in a frog's foot (p. 29). By the turn of the century some referred to motion pictures as history written in lightning. A British group coined the term *documentary* in 1936 (Wheen 1985:157).

The Soviet government and American commercial studios began to dispense information about current events via newsreels in the 1920s (Barnouw 1974) and newsreels remained a popular source of news until televised newscasts became common. Although newsreels have faded into history, recorded sounds and sights that are broadcast remain a major source of information about current affairs.

Many governmental agencies, religious groups, and commercial companies began to produce and distribute documentaries in the 1930s to advocate special interests. For example, automobile manufacturers financed the production of documentaries that glorified automobile driving. Special-interest documentaries are produced by nearly all large organizations.

Most recorded sounds and sights are made to document affairs of only limited and passing interest, but some are made to document events of extended and enduring interest. Individuals and families often produce documentaries of events in their lives that seldom are of interest to others. Other documentaries offer accounts of events of interest to many others for a considerable time. For example, the recording of President Kennedy's assassination in 1964 continues to fascinate many. Recordings, whether made for personal interest or to document an event judged to have lasting communal significance, provide a new kind of data about the past. Historians analyzing the Vietnam War can consult recordings as well as documents to acquire information, whereas the student of the U.S. Civil War can consult only static artifacts and documents.

Documentaries allow for the formulation of different conceptions of the past than those based on information provided by documents and static artifacts. For example, those who listen to and watch the British documentary *Desert Victory* form a somewhat different conception of warfare in World War II in North Africa than that formed from reading histories of warfare in North Africa based on documents and artifacts.

Whereas most people make documentaries to document particular events, some people, especially researchers, make documentaries to generate new knowledge. For example, whereas a recording might be made of a tornado to document its occurrence, meteorologists record tornadoes to accumulate data with the intent of extending our knowledge about tornadoes. Biologists, sociologists, and other researchers make recordings to obtain specimens of phenomena that they analyze to extend our understandings.

The replications provided by recordings have far greater fidelity with the phenomena they represent than the replications offered by notes, graphs, and still photography. They provide across-time replications of dynamic phenomena, whereas notes, graphs, and photographs offer point-in-time representations. Recordings of a tornado replicate the phenomena that constitute tornadoes with far greater fidelity than still photography or even still photography supplemented by memory and notes.

Audiovisual recordings provide social scientists opportunities to acquire more comprehensive representations of social phenomena than can be acquired via any other procedure (Couch 1987). For example, questionnaires call for respondents to report on phenomena, presume the respondents are competent observers of the phenomena under investigation, and provide sociological researchers with static point-in-time characterizations of social phenomena. In contrast, recordings of social transactions provide sociological researchers rich, dynamic, across-time representations of social phenomena.

Social processes are so fluid and complex that it is impossible for naked-eye-and-ear observers to reconstruct the social processes that they have seen and heard with fidelity. Recordings provide sociological researchers with far more authentic representations of social phenomena than are provided by unaided observations preserved in human memory and notes. Recordings resolve many of the observational problems for social scientists that stem from the complexity, ephemerality, and fluidity of social phenomena (Couch 1987).

Recording technologies allow sociological researchers to make more refined analyses of social phenomena than previously possible. The replay capacity of recordings allows researchers to examine specimens of social phenomena again and again, to share data with other researchers, and to systematically compare specimens of social phenomena, for example, in formulating principles of interpersonal negotiations (Sink and Couch 1986).

Audiovisual recording devices provide sociological researchers with new kinds of data that in turn allow them to formulate new approaches to the study of social processes and relationships. Just as the microscope opened new vistas for biologists, audiovisual recordings open new vistas for social scientists. The effective use of recording devices for sociological purposes, of course, requires that researchers use sociological theory to frame and guide their use of the devices.

Information technologies that preserve information in static form, such as written languages, graphs, and photographs, are more conducive to researchers adopting stable standpoints and developing abstract concepts than are information technologies that provide dynamic information. Although the dynamic data offered by recordings allow researchers to subject fluid phenomena to sustained analyses, if such research is to contribute to greater understandings the analyses must be contextualized by an abstract frame of thought and a stable standpoint must be associated with information technologies that provide static information. The merger of a reflective frame of thought based on written languages, graphs, and diagrams with the analysis of fluid data preserved by recordings offers researchers the opportunity to generate new forms of knowledge.

The spoken word allows people to provide evidence by testimony; the written word and static visuals provide evidence via documents; recordings provide evidence via documentaries. The dynamic and evocative data provided by recordings favor the formulation of across-time passionate theories of social life. In contrast, the static data provided by written languages, graphs, and photographs favor the formulation of the point-in-time dispassionate theories of social life.

FORMATTING EXPERIENCES AND ACTION

Orality, like recordings, also favored dynamic and evocative symbols, but the constancy of the information preserved orally is problematic whereas the information preserved via recordings endures. Written languages favor static and referential symbols, and with the exception of attenuation due to erosion of the material written on, written languages also favor constancy. A written description of deer frolicking on the mountainside remains more constant than an oral account or the deer themselves. An audiovisual recording of deer frolicking remains almost as constant as a written description, is as dynamic as the event recorded, and usually is more evocative than a written description.

Advocates of recording technologies claim they enrich our lives, while many of their critics advance the argument that recordings offer shallow replicas of reality and erode the ability of people to distinguish fantasy from representations of reality. Perhaps moderns are more captivated by imaginary events than their ancestors, but there is no evidence that they have greater difficulty than their ancestors in distinguishing fantasy from reality. Most people most of the time distinguish experiences derived from recordings from the experiences derived from the mundane world without great difficulty. Only the very young are likely to confuse recordings of stampeding herds of cattle with stampeding herds of cattle.

Those skilled in creating recording devices attempt to create sets of experiences for users that are isomorphic with the experiences of those who participate in and/or witness the events that are replicated. The early producers of recordings of musical performances attempted to offer the same sounds to their audiences as the sounds heard by those who attended the concert. In recent decades producers have developed techniques that allow recordings to offer experiences that cannot be obtained in the mundane world. Some techniques, such as those that magnify sights and sounds, provide experiences that are more vivid than those that can be obtained by on-site observers. The virtual realities offered by contemporary machines often never were and are unlikely ever to be. The creators of virtual reality devices attempt to encase users of the machines with a cybernetic environment that will elicit complete suspension of disbelief.

Static visual displays, oral accounts, and written narratives also offer virtual realities, but realities of less fidelity. For example, readers of electronic novels often respond as if they were participants in the activity described; they thereby migrate from the mundane world to the world of cyberspace. However, entry into cyberspace via reading is highly contingent on readers using their imaginations to project themselves into the situations described. Recordings can provide sensations that require little effort from listeners and viewers for them to enter a fantastic world in cyberspace.

Some critics of recordings claim that the dynamic evocative experiences derived from them are so engulfing that the cybernetic environments often transform listeners and viewers of recordings from reflexive actors into responders. For example, listeners of recordings of music reproduced by record players through multiple loudspeakers sometimes become so engrossed in the sounds that the only behavior they produce is unreflexive responses to the sensations they experience in the immediate present.

The activity that precedes listening to and viewing recordings is always a reflexive act For example, the donning of a Walkman© is a reflexive act. However, when attention is focused on the sensations provided by the recordings, reflexivity often recedes (Chen 1993). Those who turn to recordings for entertainment probably are more likely than those who attend live concerts to become engulfed in the sounds and sights offered. Intrusive sounds, sights, and bumps more often distract those attending a live concert than those who listen to a recording of a concert in private.

A rapid flow of intense sensations, whether derived from machines offering virtual realities or from the mundane world, tend to fragment experiences into a series of perpetual presents unconnected to either the past and future. Well-crafted recordings offer people the opportunity to be so absorbed in an ever-changing flow of sensations that consciousness of self, the future, and the past is suppressed.

However, when recordings are self-consciously analyzed, new parameters of reflectivity and reflexivity are possible. For example, when an audiovisual recording is played in reverse or slow motion, taken-for-granted sequences and durations are rendered problematic and reflectivity centered on temporal structures comes to the fore. Athletic coaches, recording specialists, and chemists are but a few who have refined and elaborated their conceptions of durations and sequences by adopting an analytic standpoint while attending to recordings. An analytic standpoint also can be adopted toward recordings of one's own action to expand one's self-consciousness.

In comparison to reading, listening to and watching recordings more often occurs in a collective context. Students commonly view a film with their classmates, but seldom read textbooks in unison. Clients of motion picture theaters are more likely to attend them with another person than are readers of novels to read in unison with another. When Americans go to a motion picture theater, over 80 percent of the time they are accompanied by at least one other person (Jowett and Linton 1980:84). However, the perfection of magnetic recording devices has been accompanied by a movement of the use of recordings from collective contexts to solitary contexts. Many don their Walkman© to block out distractions with pleasant sounds, and each day millions of people watch videocassettes of motion pictures in the privacy of their homes.

Participation in traditional ceremonies and celebrations included sharing emotions and acting in unison with others. Those who convene at motion picture theaters establish copresence and may respond in unison, but they seldom act in unison toward the sounds and sights emanating from the recordings. Those who rent a videocassette of a motion picture and watch it alone in their home do not establish copresence and neither share experiences with others nor act in unison toward the sounds and sights offered by recordings. Recorded entertainment promotes solitariness, and solitary people do not become embedded with others. Live entertainment promotes collectivism and people who emote and act with others sometimes become embedded with one another.

Paradoxically, as solitary pleasures have become more pervasive, worldwide commonality of sentiment and form of consciousness has become greater. However, commonalities derived from recordings are not equivalent to communalism based on shared experiences and collective activity. For example, participation in communal celebrations, such as singing in a church choir, generates self-other embeddedness and a communal standpoint. Listening to and viewing recordings in solitude, in contrast, contributes to narcissism and egoism.

Although the use of recordings, especially when used for entertainment, is often a solitary activity, the production of recordings is more often a collective enterprise than the production of printed documents. The production of a book, for example, typically includes interaction between authors, editors, and printers but usually the interaction is a series of dyadic encounters. In contrast, the production of a recording of a musical concert usually involves a number of people simultaneously interacting with one another.

CONCLUSIONS

The fidelity of recordings remains a marvel. The only other information technologies that approach recordings in fidelity are still photography and broadcasting. Photographs offer only static replicas of visual configurations and broadcast sounds and sights are ephemeral. Recordings offer dynamic replicas that endure.

Recordings and replications of them are so easy to make that it is difficult to control their production and distribution. Social boundaries and hierarchies based on control of the flow of written information are fading. Millions have access to camcorders and VCRs that allow them to produce and replicate entertainment and information. Millions of nonliterates acquire information and knowledge about such varied phenomena as the clothing fashions in elite salons and the engineering wonders of outer-space probes from recordings.

Pleasures that once required great expenditure of resources are now readily available to a large percentage of the world's population. Nearly all can go to a local retail outlet and select entertainment from a large number of recordings at minimal cost. People are no longer limited to accepting or rejecting traveling troubadours and relying on local talent when seeking entertainment.

Recordings have also desocialized entertainment. Pleasures that once were only available at public occasions can now be enjoyed in solitude. Recordings allow human beings to substitute solitary hedonism for solidary celebrations.

History, science, and art are being transformed as recordings compete with print for hegemony in the preservation of information and knowledge. The acuity and veracity of perception, particularly of dynamic visual events, were first enhanced by recordings of a galloping horse. That recording provided a new kind of evidence. Researchers with a multitude of different interests use recordings to accumulate and preserve evidence. Accounts of the past offered via motion pictures are assigned greater credibility by most than are assigned to history books.

The merger of recordings with computers will perhaps render books obsolete and the merger of recordings with broadcasting is rendering newspapers obsolete. Literacy may become as irrelevant as the ability to compose poems for the purpose of preserving information. Reading and writing may become an activity mastered by only a few interested in preserving skills of the past.

CHAPTER

12

Broadcasting Sounds and Sights

In 1864, Maxwell put forth the claim that the atmosphere contained electromagnetic impulses that traveled at the speed of light. He did not think the impulses had any practical value. Thirty years later, Marconi transmitted electromagnetic impulses without wires. In 1906, a procedure for transforming electronic impulses into sounds was developed, and "[b]y 1914 the world of 'amateur wireless' was a world of lively activity" (Briggs 1977:58). In 1923 there were over five hundred radio stations in the United States, with an estimated audience of two million. Within a decade after the invention of the transistor radio in 1948, radio listening achieved worldwide universality. Today more people listen to radio than read newspapers or watch television.

In the 1920s, several inventors developed procedures for transmitting sights via electromagnetic impulses. By 1940, television stations were broadcasting in Britain, the United States, Germany, and the USSR. The development of television was interrupted by World War II, but by 1948 the number of television sets in the United States increased from one hundred thousand to one million. By the 1960s several national television networks were in place and home ownership of receiving sets approached universality in North America, Western Europe, and Japan. Home ownership of receiving sets approached universality in the Soviet Union and China in the 1980s. Most owners of a television set can receive broadcasts from at least three or four stations and those with a satellite dish or cable have access to scores of television stations. In the United States, each evening about 75 percent of the population watch television.

The term *broadcasting* originally referred to the practice of planting seeds by casting them about in fields instead of planting them in rows. Radio and television stations dispense sounds and sights in an analogous manner. Or, if one wishes to be technical, broadcasting stations emit electromagnetic impulses that are transformed by receiving sets into sounds and sights.

Broadcasting is the most popular source of entertainment in most nations. Newscasts are overwhelming newspapers as a source of information about current affairs, and educational programs transmitted by radio and television stations compete with books and classrooms for hegemony in dispensing knowledge. The print media that flourish are those with a parasitical relationship with electronic information technologies. The magazine with the largest circulation in the United States is *TV Guide*.

STATE CONTROL

As radio broadcasting became a common activity, most state authorities concluded that state control was necessary. In 1912, the state authorities of Italy justified control of broadcasting by observing that the government is "an institution known for its impartiality and for its freedom from human error" (Cavazza 1979:83). In 1923, Parliament established the British Broadcasting Company (BBC), banned commercial broadcasting, and harnessed the BBC to the glorification of the empire. The authorities of the Soviet Union harnessed broadcasting to the task of creating Utopia. A few decades later the British Empire collapsed followed by the collapse of the communist dictatorship in the USSR.

Several advocates of state control of the broadcast media have argued that state regulation makes it possible to disseminate information that is uncontaminated by sentiments and interests. Such may be the case in Utopia, but there is no evidence that the information disseminated by government-operated media reflects the interests of the media operators less than the content of commercial media reflect the interests of their operators. The broadcasts of government-operated stations advocate compliance with state authorities and glorify the state elite. In a parallel manner, commercial broadcasts fan interests in everything from colored toothpicks to luxury automobiles and glorify economic institutions.

In a few nations, broadcasting was originally conceptualized by state authorities as a commercial industry that required no special regulations. However, even when broadcasting was conceptualized as just another commercial activity, problems emerged that led governments to regulate broadcasting. For example, in the United States when radio broadcasts began interfering with one another, station owners lobbied for federal regulations and a radio code was enacted in 1927 followed by the establishment of the Federal Communications Commission (FCC) a few years later.

In the early days of broadcasting some state authorities, for example those of Great Britain and Germany, promoted the development of national networks and home ownership of receiving sets. Other states, such as the USSR and China, promoted national networks linked to local loudspeakers. Radio receivers did not outnumber loudspeakers in the USSR until 1964 (de Sola Pool 1983:32), and it was a few years later before the number of radio receivers exceeded the number of loudspeakers in China. Television sets did not became household items in the USSR and China until the 1980s after those governments began subsidizing the manufacture of television sets and established national television networks.

Until recently most governments restricted "broadcasting to a 'responsible' monopoly, consisting either of government officials or especially authorized public bodies" (p. 112). However, in the past few decades broadcasting has often moved from being a governmental operation to a commercial enterprise. For example, in Britain the BBC monopoly of television broadcasting was broken in 1955 and the BBC monopoly of radio broadcasting in 1969, when domestic commercial stations were allowed to compete with it. Many nations have both governmental and commercial broadcasting stations and networks. For example, in the 1980s Thailand had four television stations: two commercial stations and two operated by the government.

Although dictatorships have been more restrictive of the broadcast media than legislature-centered states, when broadcasting has been a state monopoly in legislature-centered states, the dissemination of information about state affairs has usually been as closely controlled as in dictatorships. For example, it was not until after BBC's monopoly was broken that the broadcast media reported on the debates in Parliament. Broadcasters in legislature-centered states where stations are state operated have been only slightly more willing than broadcasters in dictatorships to broadcast information that they think might offend state authorities.

On rare occasions dictatorships have relaxed their control of state-operated broadcasting stations. In 1968, Czechoslovakian television offered "a regular one-and-a-half hour program in which people were able to challenge ministers directly in the studio" (Wheen 1985:85). Intense agitation for the overthrow of the government followed shortly thereafter. The tanks of the USSR rolled in and crushed the rebellion. Those who reinstalled the dictatorship did not try that silly experiment again.

In legislature-centered nations that did not make broadcasting a state monopoly, access to the broadcast media by political candidates emerged as a public opinion issue. The U.S. Congress attempted to resolve that issue by enacting regulations to assure all political candidates equal access to the broadcast media. In Britain, the BBC originally resolved the issue by ignoring political campaigns. Whenever dictatorships have moved toward instituting a legislature-centered state structure, access to the broadcast media has emerged as a major public opinion issue. In 1991, leaders of the opposition were allowed to broadcast on Soviet television as the state structure was transformed from a dictatorship into fragmented legislature-centered structures. The legislatures of many nations have attempted to regulate the amount of money candidates for offices can spend to acquire access to the broadcast media.

Dictators realize that if those opposing them were allowed access to the broadcast media their control of state structures would be rendered problematic. When dictators exercise control of the broadcast media, it nearly always generates cynicism toward the information disseminated by the media. Nonetheless, state control of the broadcast media stifles dissent. Even when discontent is widespread in a society, if the discontent is not disseminated via the broadcast media, organized opposition is very unlikely. Conversely, it is impossible to maintain a dictatorship if groups opposing the dictatorship have access to a nation's broadcast media. The penetration of national boundaries by the broadcast media have become so pervasive in recent decades that traditional dictatorships may be rendered obsolete.

QUALITATIVE FEATURES

The broadcast media favor the dissemination of evocative, ephemeral, and dynamic symbols. Increases in the prevalence of those types of symbols is conducive to the development of emotionality, change, and fluidity and the dilution of rationality, tradition, and rigidity. One scriptwriter noted, "Few of my scripts have been published and not many of my shows survived even in recorded form. For the rest, they are blowing in the wind indeed" (Bridson 1971:223–24).

The broadcast media transcend space, but they have no temporal depth. The only tracings that broadcast sounds and sights leave are memories, but the memories are not without significance. The sentiment, definitions of reality, and programs of action disseminated, aroused, and coalesced via broadcasting constitute a significant part of the collective consciousness of all contemporary societies.

The acquisition of entertainment and information from the broadcast media requires listeners and viewers to subordinate themselves to broadcast stations. In contrast, readers and users of recordings can subordinate these media to their wills. Readers can reread a section; users of recordings can replay a segment. Audiences of broadcasts have no control over the flow of experiences they derive from the broadcasts. Either one goes with the pacing offered by the broadcast or one drops out. Obviously, if one wishes to reflect upon or analyze broadcast information in any detail, it is necessary to record it.

Broadcast stations provide a means for the mass dissemination of entertainment and information so that those who have access to them can structure the experiences and consciousness of a multitude of others. When there is a limited number of broadcasting stations, only a few have the opportunity to structure the experiences and consciousness of audiences. The expansion of the number of radio and television broadcast stations has eroded monopolies of broadcasting outlets. Hundreds of thousands with access to broadcast stations now compete with one another to entertain and inform the world.

ENTERTAINMENT

Entertainment dominates broadcasting. Television offers a greater variety of entertainment than radio and in nearly all parts of the world television stations give more air time to entertainment than to information. In addition, most informational programs such as newscasts are designed to entertain as well as inform. Almost without exception, commercial stations unabashedly design programs to entertain as large a number as possible.

Even most state-operated stations devote the bulk of their air time to entertainment. Some claim that state-operated broadcast stations offer more uplifting entertainment programs than commercial stations. But when state-operated stations compete with commercial stations the government-operated ones have difficulty in attracting audiences. Apparently only a small percentage of the population cares to be uplifted by broadcast entertainment.

Much of the entertainment offered by radio stations provides a pleasant auditory background for listeners. Radio listeners frequently clean house, write books, or drive trucks while listening to broadcast music. Some listeners and viewers organize their daily routines so they can listen to and/or watch their favorite programs and give their undivided attention to broadcasts. But even when a broadcast is the primary focus of attention, listeners and viewers often engage in other activities. Some eat; others converse; still others groom themselves.

A substantial percentage of broadcast entertainment is composed of transmissions of sounds and sights as they occur. In the early years, the broadcast media transmitted concerts and spoken descriptions of athletic events as they were scheduled. That practice has evolved into a symbiotic relationship wherein entertaining events are scheduled to serve the broadcast media and the broadcast media render services or payment to those providing the performances. Starting times and durations of performances have been modified to accommodate the interest of broadcasters, and conflicts between the interests of those who attend the performances and the interests of the broadcast audiences tend to be resolved in favor of the broadcasters.

Although a substantial percentage of entertaining programs continue to be live broadcasts of performances, recordings have become the major source of material for entertaining broadcasts. Many programs, especially newscasts and broadcasts of athletic events, are complex mixtures of the instantaneous broadcasts of sounds and sights, commentaries on the events, and replays of recordings. Broadcasting and recording technologies are so closely merged that some conceptualize the two as a single technology.

The term *time-shifting* is commonly used to refer to the practice of broadcast audiences recording broadcasts to be viewed later, but broadcasters used recordings to time-shift broadcasts long before it became a common practice for audiences to do so and they make more extensive use of recording devices than do audiences. Recording devices partially free both broadcasters and broadcast audiences from the tyranny of schedules. Recording technologies also allow broadcasting stations to operate at less cost.

A few decades ago nearly all American communities celebrated Memorial Day with parades, speeches, and entertaining performances offered by local talent. Only a small percentage of American communities currently hold Memorial Day celebrations. Instead millions watch the television broadcast of the Indianapolis 500. Many homes in the wealthier nations have several radios and two or three televisions. In such homes each family member can select broadcasts on the basis of personal preference and acquire pleasure in isolation. Entertainment thus has become less communal as the broadcast media have matured.

Much of the gratification derived from temple-centered celebrations stems from the mutuality and solidarity generated among the participants; in contrast, the gratification acquired from broadcast entertainment is derived from sensations provided by the broadcast sounds and sights. Many television broadcasts provide a host of visual delights. The images offered often change every few seconds. The newscasts of most commercial stations are an ever-changing kaleidoscope of sights and sounds.

Broadcast entertainment often transforms conceptions of the past and sometimes provides a foundation for the creation of new social enterprises. For example, the television series "Roots" offered an account of the past different from that offered by written histories. Subsequent to the broadcasting of "Roots," two hundred colleges and universities in the United States added African-American courses to their curriculum. The consciousness of slavery that prevailed in the United States prior to the maturation of the broadcast media and recording technologies reflected the bias of the written word; current conceptions reflect the biases of broadcasting and recording technologies. The distinction between entertainment and information has always had a large twilight zone, but that twilight zone has become much wider with the development of broadcasting and recording technologies.

INSTANTANEOUS NEWS

The development of techniques for the transmission of information about events as they occur has transformed the meaning of news. As the Earth has become encased in a grid of telegraph networks, what happened prior to yesterday is no longer news. The development of satellite broadcasting, international telematic networks, and the universality of ownership of radios and televisions has been accompanied by the development of the conception that only what is happening in the immediate present qualifies as news. What happened yesterday, as it were, has become ancient history.

That transformation began in the 1920s and was brought to fruition with the establishment of continuous news networks in the 1980s. The results of the national election were broadcast in the United States as they became known in 1924 for the first time. In 1940, descriptions of aerial battles between the RAF and the German Air Force were broadcast by BBC as they occurred. The live broadcast of the first American moon landing in 1969 attracted an estimated audience of 723 million around the world (Wheen 1985:233). In 1991, a worldwide audience listened to and watched the bombing of Baghdad.

As newscasts became commonplace, people came to regard broadcast news as more valid than printed news. In 1939, a majority of U.S. citizens indicated that they regarded radio news as more accurate than that provided by newspapers (Czitrom 1982:86). Television surpassed newspapers as the reported choice as source of news in 1967 in the United States. In the early 1980s, over 80 percent of the adults in the USSR watched televised news (Mickiewicz 1988). Prior to the perfection of broadcast technologies, informed citizens read newspapers, but in the broadcast age they watch televised news.

Instantaneous news has partially erased the dichotomy between witness and nonwitness. Millions claim to have witnessed the crushing of the student rebellion in China and the overthrow of Romania's dictator in 1989. Television audiences may acquire more information more quickly about a newsworthy event than those participating in events themselves.

The multiple angles of coverage, close-ups, replay capacity, and split screens of television broadcasts often provide more information than is available to on-the-scene witnesses. Some sports fans bring portable televisions to athletic events to enhance their perceptions of the events.

Prior to the maturation of the broadcast media, information about international affairs nearly always reached heads of state before it reached rank-and-file citizens. In contrast, however, during the 1991 Gulf War, the director of the CIA telephoned the president of the United States to inform him that Iraq had launched a SCUD missile attack on Israel. The president replied that he was watching the attack on television. Awareness of the latest developments at a meeting of the United Nations enters the consciousness of the politically informed residents of Tokyo almost as rapidly as it enters the consciousness of on-the-scene participants. The broadcasting of instantaneous news deprives state authorities of a time differential between when they learn of current events and when ordinary citizens learn of them. The loss of that differential sometimes pressures authorities to act more rapidly than they might have when news was disseminated via the telegraph and newspapers. However, the significance of instantaneous news for state affairs is suggested by the practice of many high-ranking officials surrounding themselves with televisions turned to instantaneous news. One Pentagon official asserted that the best information he had about the success of the bombing of Iraq in 1991 was provided by the newscasts of a continuous news network.

When newspapers were the primary source of information about current affairs, editors and reporters exercised greater control over the flow of news than over the sources of the news. That ratio has been reversed as the broadcast media have become the primary conveyor of news. The content of newscasts is determined by the sources of news more frequently in the broadcast age than in the newspaper age. Even when news is selected and processed by professional reporters, broadcast news has distinctive qualities. Television reporters are interested in acquiring sight and sound bites of dramatic events, whereas newspaper reporters are more likely to seek information about behind-the-scenes activities and relationships. The dissemination of information about behind the scene activities and relationships increases the accountability of state officials; on the other hand, sight and sound bites of the dramatic actions provide officials a stage they can use to elicit appreciation from audiences.

The evocative bias of broadcast news provides producers of newscasts greater opportunities to generate compassion from audiences than print provided newspaper editors and reporters. A newscaster who wishes to elicit sympathy for the plight of war refugees can broadcast dynamic close-ups of suffering people. Newspaper accounts of the same disaster may include photographs but, in general, still photographs are less evocative than sound and sight bites. Television newscasts, especially commercially sponsored newscasts, often are extravagant collages of very short evocative sound and sight bites.

In many instances in legislature-centered states, controversies are initiated by the investigative reporting of newspaper reporters and then transformed into television extravaganzas. For example, newspaper reporters initiated news of the Watergate scandal. After the scandal had achieved considerable currency as a consequence of newspaper accounts, television stations and networks began to report on the affair. These reports were dominated by on-camera appearances of the principals. In contrast, newspaper coverage continued to focus on who had contact with whom and who was responsible for what action.

The distinction between newsworthy events and reports about them has become ambiguous as newscasts have replaced newspapers as the primary source of information about current affairs. When newspapers were the primary source of information, reporters usually were appendages of the events themselves. In contrast, television reporters, anchorpersons, and producers frequently are active participants in newscasts. For example, interviews of heads of states by reporters of the broadcast media are often broadcast live, with reporters sharing center stage with heads of state before a worldwide audience. According to some, the broadcast media have brought the age of journalism to an end (Altheide and Snow 1991).

ELECTRONIC CHARISMA

The state officials of Nazi Germany were the first to mount an organized effort to use the broadcast media to create a state structure that rested in large part on emotionally bonding citizens to the head of state. A Nazi directive of 1934 stated, "[B]roadcasting will be used to unite the entire nation" (Kris and Speier 1944:52). After becoming the head of state, Hitler routinely used radio broadcasts to offer definitions of reality to a national audience. Local radio wardens noted the quality of reception, encouraged people to listen with family, friends, and neighbors, and reported those who listened to foreign broadcasts to the authorities. When Hitler addressed the nation, evocative symbols were accentuated. Marching music was played at the beginning and end of each broadcast (Hafstaengl 1957) and Hitler always addressed a live audience (Kris and Speier 1944:125). The radio audience thus heard the responses of an immediate audience as well as Hitler's speech.

The infusion of state structures with a charismatic relationship focused on the head of state, however, has not been limited to dictatorships. In the United States, Franklin Roosevelt delivered fireside chats in the 1930s to elicit support for his definitions of reality. In contrast to those of Hitler's, the introductions to Roosevelt's addresses were brief, he made less use of evocative symbols, and listening to foreign broadcasts was not prohibited. But, the basic strategy was similar, and Roosevelt was able to emotionally bond a significant percentage of American citizens to his personality. When he requested listeners to write and tell him about their troubles, he received half a million letters (Minow, Martin, and Mitchell 1973:30).

Even in Britain during World War II, where a policy of keeping politicians off the airways prevailed, the state structure became infused with a charismatic relationship focused on the head of state. On September 1, 1939, the prime minister informed British citizens via radio that they were at war with Germany "before he went down to the House of Commons to tell their representatives" (p. 4). In a similar manner, on December 8, 1941, Roosevelt's request that Congress declare war on Japan and Germany was broadcast live on national networks.

Prior to World War II, the terms *charisma* and *charismatic*, derivatives from a Greek word that meant "touched with grace," had currency only among scholars who used them to refer to social upheavals wherein traditional structures were rendered problematic by leaders of revolutionary movements by establishing an emotional bond between themselves and their followers. Those terms entered vernacular languages, though, as heads of states began to use radio broadcasts to generate an emotional bond between themselves and citizens.

Traditional charismatic relationships begin with one person acquiring a reputation for transforming common dissatisfactions into shared dissatisfactions and articulating a utopian future. After a following has been achieved, gatherings are held to validate and extend the charismatic relationship. At these gatherings, the leader offers monolithic definitions of reality, describes utopian futures, and generates shared emotions (Couch 1989). The members of the audience communicate to the speaker and with one another largely by evocative symbols, while the elevated person uses both evocative and referential symbols to define reality and program the future. The speaker speaks and the audience laughs, shouts, claps, and moans in unison.

Often the state structures of a nation are subtly infused with a charismatic relationship focused on the head of state. A prototypic instance of that is President Kennedy's, "And so my fellow Americans ask not what your country can do for you—ask what you can do for your country." When such statements are broadcast, the call is for all to subordinate their interests to the interests of the state as articulated by the head of the state. Hitler issued the same type of appeal when he called on the German citizens to resist the allied invaders.

Of course, not all heads of state who have attempted to infuse a state structure with a charismatic relationship have been successful. In 1964, de Gaulle's opposition was supported by the press but he had control of the television network. While he accordingly expected electoral success, he nonetheless failed (Bagdikian 1971:15). In Togo and Kenya the presidents dominated television broadcasts (Ganley and Ganley 1987:8) but failed to infuse those state structures with a charismatic relationship.

Even when a head of state has effectively used the broadcast media to infuse a state structure with a charismatic relationship, sometimes his or her elevated position has been rendered problematic. Ceausescu, the dictator of Romania, had infused the state structure of Romania with a charismatic relationship, but many Romanians became disenchanted with him. On December 21, 1989, he attempted to validate his elevated position by speaking to an immediate audience and broadcasting his speech over a nationwide television network. The immediate audience booed him, he became frightened, and the broadcast was abruptly terminated. A citizen living in another city noted, "Then we knew he was weak and scared. The whole country saw him frightened. This was a very important moment" (*New York Times,* January 5, 1990). Two days later the insurgents informed the world via television that Ceausescu and his wife had been put to death.

When heads of state use national networks to communicate to citizens, they collapse state structures into two layers. For the duration of the broadcast, all members of the audiences are placed in an equal and subordinate position. If no others have access to the mass media, only the definitions of reality offered by the head of state are widely disseminated. When heads of state have a monopoly of access to the broadcast media, it provides them with the opportunity to lie on a grand scale. When Hitler asserted via a radio broadcast to the German citizens in 1939 that the Poles had initiated the war between Germany and Poland, very few German citizens had the resources to assess the validity of the claim. Likewise, when Nixon asserted on a nationwide television broadcast that no American soldiers had invaded Laos in 1971, his definition of reality could not be effectively challenged by many citizens.

Not all citizens automatically accept the definitions of reality offered by heads of state. Even in nations such as Nazi Germany, a substantial proportion of the citizens refused to attend to the broadcast speeches of heads of state. However, unless those offering alternative definitions of reality have access to the broadcast media, the definitions of reality offered by heads of state are the only ones of significance.

In traditional palace-centered nations, coronations, state-sponsored festivals, and national celebrations affirmed state structures. In traditional legislature-centered nations, the print media, especially newspapers, disseminated information about state affairs and the state structure was affirmed by voting, public opinion, and national celebrations. When a state structure rests on a foundation of a charismatic relationship, state structures are affirmed by a broadcast speech of the head of state. When the attempt by the German generals to assassinate Hitler in July 1944 was crushed, for example, Hitler immediately addressed the nation via radio to affirm his dictatorship.

Traditional dictatorships primarily rested on a foundation of terror and monopolies of knowledge. Hitler, Khomeini, and Ceausescu also relied on terror and monopolies of knowledge, but in addition they used the broadcast media to generate gratitude from citizens. In those states unity stemmed from a combination of common sentiments elicited by broadcasts and conformity solicited by threat of violence. State structures that rest on a combination of charisma and terror involve many citizens who willingly serve as instruments of the state.

In the early decades of broadcasting, many nations were blanketed by a single network or at most two or three networks, which facilitated the efforts of heads of states to form a charismatic relationship with citizens. The proliferation of satellite broadcasting, cable networks, and videocassettes, however, has made it difficult for state authorities to control the flow of broadcasts within their borders and that lessens their ability to infuse state structures with a charismatic relationship. However, those developments may be counterbalanced by the movement of news from newspapers to newscasts. Newscasts provide heads of state with more opportunities to infuse state structures with a charismatic relationship than newspapers. National celebrations such as the Fourth of July in the United States and Bastille Day in France are commonly broadcast on a national network, and typically a statement by the head of state is a part of the broadcast. In the prebroadcast age, such ceremonies were perhaps attended by tens of thousands, and newspapers provided accounts for those not in attendance. Those accounts typically were much less evocative than live broadcasts.

Appearances of and statements by heads of state compose a significant part of most newscasts. For example, in the 1980s the heads of state of the United States and USSR accounted for between 5 and 10 percent of the time devoted to showing people on the national newscasts of their respective nations (Mickiewicz 1988:143). Those appearances facilitate the construction of charismatic relationships focused on heads of state.

In legislature-centered states, the relationships between citizens and state officials are characterized by a multilateral flow of dictates and accountability: Constituents dictate to representatives and representatives are accountable to citizens; representatives dictate to authorities and authorities are accountable to representatives; and state authorities dictate to citizens and citizens are accountable to authorities. In contrast, in nations where the state structure is infused with a charismatic relationship, there tends to be a unilateral flow of dictates from heads of states to citizens; citizens are accountable to state officials, but officials are only minimally accountable to citizens.

LEGISLATURES AND BROADCASTING

Some critics claim that the broadcast media do not provide the information about state affairs necessary to maintain legislature-centered state structures. For example, Epstein (1973) claims that television news is news from nowhere and Postman (1985) develops the theme that the broadcast media are eroding democratic structures by enticing us to amuse ourselves to death. An event commonly proffered to demonstrate that the broadcast media have lessened reflective assessments of candidates and officials is the nationwide television address by Richard Nixon in 1952. Nixon had been accused of illegal activity, but he saved his candidacy with an evocative broadcast in which among other things he made emotional references to the family dog. The avalanche of letters, telephone calls, and telegrams to Republican headquarters following his speech ran 350 to 1 in favor of keeping Nixon on the ticket (Mickelson 1989:56). Nearly all contemporary candidates for elective office now emphasize evocative symbols to solicit appreciative responses from voters. The success of modern candidates has become so contingent on generating at least a fleeting charismatic relationship between themselves and voters that it has become almost impossible for a person not widely viewed as attractive to be elected to a national office.

As the broadcast media became the primary source of information about current affairs, membership in political parties has declined. In 1960, 42 percent of the adult citizens of the U.S identified themselves as a member of a political party but only 23 percent did so in 1972. Participation in voluntary organizations declined in the USSR as television blanketed the nation in the 1980s (Mickiewicz 1988:210). Furthermore, even among those voters who belong to a political party, party membership has had less impact on voting behavior than it did in the prebroadcast age.

Several legislatures have enacted laws that restrict the use of the broadcast media by heads of state to lessen the likelihood of heads of state over powering the legislature. In addition, a few legislatures have attempted to empower themselves by using the broadcast media to communicate to constituencies. For example, in 1979 the U.S. Congress established a national television channel to broadcast information about legislative affairs. Perhaps legislatures will become more effective vis-à-vis heads of state as they learn to use the broadcast media, but to date it seems the maturation of the broadcast media has been accompanied by a decline in the vitality of legislatures and the enhancement of the influence of heads of states.

Broadcasters, heads of state, and public relations specialists have re-placed reporters, editors, political parties, and legislators as agenda set-ters in many democratic nations as sound and sight bites have replaced headlines and editorials. The resolution of public issues has moved from backroom negotiations by professional politicians to media blitzes and public opinion polling. Some advocates of the broadcast media note that the broadcast media provide citizens with the opportunity to observe their representatives negotiating, grilling one another, and questioning witnesses and thereby extend democratic processes.

Such seems to have been the case in at least a few instances. For example, after years of military dictatorship, Brazil elected a president in 1989 who was impeached for corruption shortly after he was elected. Similar sequences of events have occurred in many other nations and nearly always they were followed by a military dictatorship reassuming control of state structures. That did not happen in Brazil in 1989. On the day of the vote for impeachment, it is estimated that 90 percent of the adults watched some or all of the television coverage and the president was replaced without military intervention in the political structure.

Some advocates of the broadcast media have advanced the argument that the broadcast media have contributed to the transformation of dic-tatorships into democracies. It has been noted that even in the USSR where the broadcast media were state controlled, television broadcasts increased the flow of information about state affairs and the collapse of the Soviet dictatorship occurred within a few years after a national televi-sion network had been put in place. However, it was also the case that in the preceding decades there had been a large increase in the literacy level in the USSR, which was accompanied by an increased awareness of alter-native forms of state structures. If the broadcast media do in fact contrib-ute to the supplanting of dictatorships with legislature-centered states, it seems that the information that contributed to that transformation proba-bly does not stem from domestic broadcasts per se, but from broadcasts that penetrate national boundaries.

THE INTERNATIONALIZATION OF BROADCASTING

In the early years of broadcasting, several governments prohibited listening to foreign broadcasts. For decades it was illegal for British and Soviet citizens to listen to foreign radio broadcasts, and during World

War II residents of Germany who listened to foreign broadcasts ran the risk of the death penalty. Some state authorities continue to attempt to control the listening and viewing behavior of their citizens, but their efforts are relatively ineffective.

In the world of broadcasting, boundaries tend to be equated with broadcasting range. Commercial radio stations led the way in broadcasting across national boundaries (Saerchinger 1938) but state-operated stations were not far behind. Prior to and during World War II, several nations, including some that prohibited their citizens from listening to foreign broadcasts, directed broadcasts to foreign audiences. In the 1980s, the Voice of America was broadcast from 123 stations in thirty-five languages and claimed a daily audience of twenty-five million, and the BBC broadcast in thirty-nine languages to a worldwide audience (Ganley and Ganley 1989). There is little reliable evidence that violations of national boundaries by the broadcast media have undermined the state structures of nations, but they certainly have increased the awareness of international affairs for millions on a worldwide scale and thus may contribute to domestic turmoil, as they did prior to the collapse of the communist regime in Poland when between ten and twenty million Poles listened to the BBC, Radio Free Europe, and Voice of America (Katsh 1989:156).

National boundaries are effectively altered also by commercial corporations that establish pirate radio broadcasting stations and beam their broadcasts into nations that ban commercial broadcasting. One pirate station anchored in the English Channel, for instance, had a listenership of seven million in Britain within three weeks after it began operations (Harris 1977:21). In response, the prime minister of Britain instituted legislative action to enforce the prohibitions against British citizens listening to pirate broadcasts, but in the face of popular opposition he withdrew his proposal (p. 28).

Satellite transmissions of television broadcasts have had enormous effects on boundaries. They began between Europe and North America in 1962, and by 1968 were worldwide. INTELSAT III became operational in 1969, and by 1980 "more than 135 nations were using Intelsat system service full time" (Larson 1984:2). Satellite broadcasting allows all with a satellite dish or a cable hookup to instantaneously receive television broadcasts that originate on the other side of the earth.

In 1982, authorities in the USSR asserted that they would destroy any satellite that transmitted unauthorized broadcasts into Soviet territory. As recently as 1987, the USSR jammed Voice of America broadcasts. During the 1991 Gulf War, most Iraqi citizens listened to foreign radio broadcasts to obtain news about the war despite the threat of the death penalty for doing so. Several Muslim nations continue to prohibit their residents from watching television broadcasts of foreign origin, but many wealthy citizens have installed satellite dishes and the governments have not diligently enforced the bans against them. On many occasions fundamentalist neighbors of those with satellite dishes have cut the cables to the satellite dishes, shot at the dishes, and spoken out against the immorality of the owners of the dishes. Despite these and other efforts to slow the spread of satellite dishes, their popularity continues to increase around the world, and efforts to control the listening behavior of residents have been discontinued in most nations.

CONCLUSIONS

Broadcast sounds and sights have become part of daily life for nearly everyone. The airways are almost literally filled with informing and entertaining sounds and sights, and most people have access to at least some of the programs that are broadcast. The broadcast media have extended our horizons. Many holidays have been transformed from local celebrations to global extravaganzas.

In the preprint age, communalism was localized and based in large part on rituals and ceremonies wherein priests and kings communicated to proximal audiences. Print, especially newspapers, introduced a communalism based on written languages and contributed to nationalism. Both forms of communalism continue to endure, but the broadcast media have introduced a communalism based on dynamic sights and sounds that encircle the Earth. The communalism created via broadcasting is infused with greater emotionalism than communalism based on print.

Bureaucratic structures that were adjuncts of palace-centered state structures provided the foundation for the first nations. Written records linked to centralized coercion allowed warlords and bureaucrats to bring large territories under a centralized sovereignty. Printed accounts of the affairs of state distributed by commercial printers provided a cornerstone for legislature-centered states. The transformations in state structures accompanying the maturation of instantaneous news may be as profound as the dethroning of monarchs that accompanied the maturation of commercial newspapers. Televised news may render newspapers as obsolete as newspapers rendered town criers, and representative democracy may go the way of monarchy.

The use of the broadcast media by heads of state to infuse state structures with a charismatic relationship may be limited to a particular developmental phase of the broadcast media. However, it appears that representative relationships are undergoing significant modifications toward greater passion as political processes have moved from print to broadcasting. The continuation of legislature-centered states may be problematic. But totalitarian state structures probably are not the only alternative to legislature-centered ones. The substantial increase in the amount of information about current affairs dispersed that accompanied the maturation of the broadcasting media and the increase in evocative symbols may be sounding the death knell of both totalitarian and legislature-centered states.

Conversely, however, it has become commonplace for those with access to the broadcast media to provide information and gratification to a multitude of people and thereby create highly asymmetrical structures. Those with access to broadcast media compete with one another for hegemony in establishing moralities. Both entertainers and heads of state appear on television and attempt to persuade audiences to adopt their viewpoints. Entertainers become heads of state and heads of state become entertainers. When watching television it is often impossible to determine if one is observing a fictive production or the interworkings of the state.

International relationships are no longer limited to intergroup negotiations, trade, and warfare. Performers and reporters compete with the state elite for the status of prima donna on the international stage. The commercializing of the broadcast media has elevated the owners of networks to the status of kingmakers. It is they who control who performs on the international stage. One apocryphal tale has it that Ted Turner declined the opportunity to be a candidate for president of the United States on the grounds that he did not wish to lessen his ability to influence world affairs.

13

Information-Processing Machines

Computers were originally designed to process quantitative information, but have evolved into machines that also store information, process text and graphics, and receive and transmit information. Modern electronic information-processing machines have about the same relationship to the early computers as phonetic writing has to ice age depictions. The French term *l'ordinateur,* which originally referred to a device for ordering threads in weaving, more accurately characterizes information-processing machines than does the term *computer.* But it is unlikely that the term *computer* will be dislodged in the near future among English speakers as the common term for information-processing machines.

Except when merged with recording technologies, computers have not had a great impact on the quality of the information preserved and dispensed. The information processed by computers has much the same qualities as the information that was previously preserved and dispensed through written languages and print. The claim that computers have ushered in a distinct information age is as faulty or correct as the claim that the printing press ushered in a new information age. That is, the ratio of the increase in the amount of information preserved and circulated in Western nations in the last fifty years compares favorably with the ratio of increase in information preserved and circulated in the two or three centuries following the development of the printing press in Europe. The qualitative transformations of social structures that accompany the maturation of computers are a consequence of changes in the amount of information processed, not of qualitative changes wrought by processing information through computers.

Computers provide human beings with the capacity to manipulate vast quantities of information in extremely complex ways with great accuracy and rapidity. Those who use computers to manage quantitative information can add, subtract, multiply, and divide so much more rapidly than can be done manually that the two activities seem to have little in common. In a similar manner, when computers are used as word processors, information can be manipulated so rapidly and in such complex ways that the activity seems to have little in common with handwriting. Word processors offer a combination of the advantages offered by typewriters and printing presses, and they reduce problems of producing documents because of a lack of manual dexterity.

Information-processing machines have become almost as ubiquitous as television sets. In wealthier nations most students from grammar school through graduate school have access to word-processors. Nearly all merchants, even those selling out of tiny shops, use hand-held computers to process information. Computers are such an integral part of most complex organizations that when the central computer of an organization goes down, all large-scale coordinated activity ceases and all attention turns to bringing the computer back to operational status.

EARLY DEVELOPMENTS

Pen and paper computational procedures and Arabic numerals almost entirely replaced tallies, reckoning boards, and Roman numerals in Europe during the nineteenth century. Simultaneously, state bureaucracies, commercial enterprises, and universities were processing larger quantities of quantitative information than ever before. The abacus, while popular in Asia, was not widely used in Europe. The slide rule, invented early in the seventeenth century, was widely used to process quantitative information. The popularity of the slide rule was so great that most engineers carried one attached to their belt while at work, until the transistor computer rendered slide rules obsolete in the 1950s.

A category of specialists in computing—called computers—emerged in Europe during the nineteenth century, but were extremely error-prone. Babbage designed a mechanical computer that incorporated techniques developed to mass produce cloth. Babbage's computer was steam powered and, hypothetically at least, could be programmed to complete a series of computations wherein later computations were informed by earlier ones. That is, it had a memory. Despite Babbage's innovations though, the machine was not a practical success.

Several adding machines were invented in the latter part of the nineteenth century. The first one to achieve practical success, called an arithmometer, was developed in 1891 (Weinberg 1990:19). At about the same time, Hollerith, an employee of the U.S. census bureau, used punch cards and mechanical card-sorting devices to process census data. Later, Hollerith formed the Tabulating Machine Company, which subsequently become the first commercial giant in the computer industry: International Business Machines (IBM).

The foundation for electronic computers was provided by Atanasoff's electronic digital computer (Burks and Burks 1988:1). That machine matched the on-off property of electrical current and the presence or absence of a hole in a punch card with a base-2 place value system. The age of the electronic computer was ushered in when the ENIAC was unveiled in 1946. The replacement of vacuum tubes with transistors began in 1948, and silicon chips began to replace transistors in 1971.

The silicon chip and associated developments vastly increased the storage capacities of computers and made it possible to immediately interlink all information retained in a computer. Entrepreneurs began selling computer kits to computer hobbyists shortly after the silicon chip was invented. "A score of personal computer companies quickly followed" (Weinberg 1990:30). In four decades, computers changed from in-house stand-alone giants (ENIAC weighed thirty tons and required continuous maintenance) capable of managing only quantitative information to desktop machines capable of managing text and graphics as well as quantitative information that could be linked to telecommunication systems. As recently as the late 1970s, about 90 percent of computer use was to process quantitative information, but ten years later 90 percent of computer use was to process text and graphics (Batty 1987:192).

In the 1960s most computers were used by employees of large organizations. "When the first mammoth computers were built, the belief prevailed that no commercial market existed for the services that such behemoths could provides" (Flamm 1988:205). "As late as 1968, there were only 50,000 installed computers in the United States, 3,000 in Great Britain, and 5,000 in West Germany" (Weinberg 1990:61). In the 1970s commercial companies began to sell computers to individuals. The term *software* was introduced into the English language in the 1960s and entered the vernacular in the 1980s as computers became household items in the wealthier nations. In 1983 *Time* selected the computer as the man of the year. In 1984, the retail value of desktop computers surpassed that of mainframe computers in the United States.

The decline in the cost of computers has averaged about 20 to 25 percent a year for the past three decades. Their "prices have fallen at a steady rate that eclipses any sustained price decline in recorded history" (Flamm 1988:1). A small computer costing one hundred dollars can process quantitative information as rapidly as could one costing one million dollars in the late 1960s. Operating costs have declined at a corresponding rate. A computation that would have cost thirty thousand dollars in 1950 could be produced in the late 1980s for a dollar. In the early 1970s it cost about twenty-five dollars to store one million bits of information, and ten years later the cost had dropped to fifty cents.

THE STATE, MARKET, AND COMPUTERS

All of the early developers of computers received support from the state. Babbage received financial grants from the British government, Atanasoff was a professor at a state university, Hollerith was employed by the census bureau, and Mauchly and Eckert, who invented ENIAC, were supported by the U.S. Department of Defense. As late as 1959, about 85 percent of the developmental research cost on computers in the United States was supported by the federal government (Flamm 1988:16). Most of the early refinements and elaborations were made at computer centers that were adjuncts of universities. Subsequently commercial manufacturers of computers created research and development departments that competed and cooperated with universities and governmental agencies to perfect computers.

Until after World War II, Britain competed with the United States in perfecting information-processing machines, but Britain failed to develop a vigorous commercial computer industry (p. 167). In 1957, the Japanese legislature targeted electronics as a national priority (p. 178) and Japan emerged as a leading producer of computers. As in the United States, much of the innovative work on computers in Japan was completed at universities and supported by the state, but not all the innovations were developed by university employees.

Many of the innovations that transformed computers from gigantic machines into household items stemmed from informal clusters of electronic buffs, known as *hackers*, who formed parasitical relationships with the computer centers. The hackers "had no respect for power other than the power that someone could exert over the computer" (Turkle 1984:203), and most of them refused to become part of the structures in which computers were embedded. Despite their marginal position, or perhaps because of it, hackers led the way in transforming computers from machines that could only store and process quantitative information into machines that also process narrative and graphic information and visually display the information on electronic screens.

The boundaries between the hackers and professionals were and remain fluid. In the early years most professionals worked on perfecting the processing of quantitative information. Some hackers also contributed to improvements in the management of quantitative information, but most were more intrigued with the possibility of transforming quantitative information into dynamic visuals and sounds. Hackers were the first to link electronically processed quantitative information to electronically displayed configurations. For example, they invented the first computer game: Star Wars. A few hackers became entrepreneurs; others obtained faculty appointments at universities. Some continue in a parasitic relationship with computer centers.

The relationship between state structures and computer development continues to be highly varied as state authorities, universities, research institutes, and computer specialists debate and negotiate their relations with one another. In France, the merging of computers with telecommunication systems proceeded under state sponsorship, whereas in the United States that merger was effected by commercial interests. The authorities of several nations have assigned high priority to computer development, while others have been indifferent to it. Whatever the stance of governmental authorities toward computers, however, the maturation of computer technologies has been accompanied by changes in social structures, has transformed interorganizational linkages, and has changed the linkages among types of information technologies.

THE NATURE OF THE BEAST

In the early years the mass media popularized the belief that computers were superhuman thinking machines. The computer was referred to as "the brain" and television as the "boob tube." The awesome response to computers has faded, though, as manufacturers have promoted computers as user-friendly machines. That metaphoric change in part has resulted from a change in machine use and function; that is, from numeric computation functions to those of text and graphics. The issue of whether computers "think" still exists though, with some insisting that they can (Moravec 1988), others arguing that thinking machines will soon be constructed (Feigenbaum and McCorduck 1983), and still others regarding the claim that a machine can think as absurd (Weizenbaum 1976). The controversy is partially a question of semantics. If the processing of information is accepted as thinking, then computers think.

Other information technologies are designed to preserve and/or disseminate information, but computers are designed to process information. When people use other information technologies, the information is modified but the other technologies are not designed to transform information. In contrast, computers are designed to transform information.

The transformations are achieved by linking electronic impulses to numeric concepts. Only those who are expert in electronics and sophisticated in numeric concepts can program the transformation processes. Most computer users only understand the output of computer processes, not the processes themselves, and some are awed by computerized processes.

The awesome response of some toward information-processing machines stems in part from the fact the movements of machines that transform matter usually are readily observable, whereas the movements of machines that transform information usually are not. For example, the movements and impact of a trenching machine are as observable as the actions and impact of a person digging with a shovel. In contrast, the processing of information by computers, like human thought, is not readily observable to most of us. The human action (thinking) that solves a numeric equation can be entirely covert; only the result of thought processes may be observable. In a similar manner, for most users, computers covertly transform information.

Computers do not formulate intentions nor do they assess their movement toward objectives. Computers are as incapable as trenching machines of assessing ongoing activity by relating the present to an intention. Human beings, when digging a ditch with a shovel or with a trenching machine, assess the current state of affairs by relating the present to the intention of digging a ditch.

Some who anthropomorphize computers note that the memory of computers is similar to human memory. It is true that some of the operations of computers are informed by programs and previously completed operations. But the memory of computers, like that of recordings, is complete (except for attenuation due to erosion of the material) and not creative. In contrast, human memory is selective and sometimes creative.

The memories of human beings include evocative symbols as well as referential ones. Experiences that have no affectivity do not become a part of human memory. All items entered in computers are retained without affect. If human beings desire to, most of them can hold their emotions in abeyance while processing information. When they do that, they process information without displaying affect, but affect nonetheless informs their processing of information.

The symbols and concepts of human beings have referents. In contrast, the symbols and concepts processed by computers are senseless; they have no referents for the computers. Only disembodied symbols inform the operations of computers. All symbols, including the most abstract concepts, that are processed by human beings rest on a foundation of external referents.

FORMATTING EXPERIENCE AND ACTION

Some who spend much of their time using computers become so involved in computer formats that they characterize and assess their actions in terms derived from computer formats. "A computer scientist says, 'my next lecture is hardwired,' meaning he can deliver it without thinking" (Turkle 1984:17). Psychotherapy is referred to by some as "debugging personalities."

Computers allow people to transform information without reflecting about the processes that culminate in the transformations. Those with access to a computer need not multiply numbers in their head or with paper and pencil. All they have to do to solve a problem is to enter the numbers in a computer and press the proper function key. The loss of the ability to mentally compute information that accompanies the extensive use of computers is similar to the loss of the ability to memorize large quantities of information in poetic form that occurred when written languages replaced orality.

The original computers processed only static and enduring information, but the information was far more malleable than printed information. For example, authors using a word processor can revise manuscripts much easier than can authors using a typewriter. When computers are linked to recorders, dynamic information as well as static information can be rapidly manipulated in many complex ways. Those who make extensive use of computers tend to regard all information as malleable; for computer specialists no information is etched in stone.

Whether persons manipulate static or dynamic information with computers, their actions tend to be future-centered. When a person uses a computer to analyze data or to play a video game, a future is projected, information is manipulated, and the outcome is assessed in terms of the projected future. Computer use always includes the continual assessments of the present in terms of intentions. That activity is a more fluid form of analytical thinking than analytical thinking based on information preserved in traditional print. Computers are giving the concept of rationality new parameters.

The changes in how information is formatted that have accompanied the maturation of computers may be more radical than those associated with other information technologies. But the transformations in how information was formatted that occurred when written prose replaced poetry seem equally severe. However, the replacement of orality as the dominant technique for preserving information with writing spanned millennia, whereas computers have supplanted manual computations in a

few decades, and computers in conjunction with recordings and broadcasting also have radically transformed the preservation and dissemination of dynamic as well as static information.

As literacy became universal, written contracts and signatures replaced
handshakes and other interpersonal gestures as the dominant procedure
for validating commitments. As literacy approached universality, agreements put in writing became the ultimate commitment. In a parallel manner, as computers have become common, the construction of commitments
has moved from written contracts to making an entry in a computer. Many
people have been informed, for instance, that they were without hotel
reservations on the grounds that no record of the reservation was in the
computer.

COMPUTERIZED SOCIAL STRUCTURES

Computers first acquired recognition as a significant information technology by becoming an integral dimension of large organizations. In
subsequent years, computers and computer networks have become integral dimensions of nearly all organizations populated by more than a few
dozen people. In 1950, the U.S. federal government had two computers, in
1960 the number had grown to 5,277 (Westin and Baker 1972:29). Organizations that preserve information in file cabinets have become almost as
obsolete as South Sea navigators who preserved navigational lore in poetic form.

Information is almost literally at the fingertips of all modern administrators. For example, the administrators of computerized hospitals have
almost immediate access to more information about patients than hospital
administrators of a few decades ago imagined possible. The information
stored in a hospital computer may include a history of a patient's ills and
injuries, age and sex of siblings, pulse rate when the patient was on the
operating table, how the last hospital bill was paid, weight gained or lost
while in the hospital, attending physicians, x-rays, number of visitors,
and a host of other details. Should a patient enter another hospital and an
administrator of that hospital learn of the patient's previous hospitaliza-
tion, the patient's file can be readily transmitted to the second hospital.

Computers allow organizations to deliver superior services than before. For example, computer-assisted diagnoses allow physicians to apply
more medical knowledge when treating patients than is possible when
they rely only on their memories and books. Computers are less likely to
"forget" to take into account symptoms than is a physician who relies on
her memory. Computers also enhance medical treatment. For example,

computerized prosthetic limbs allow for control of the limbs by small muscle contractions. Computers are also used to monitor vital signs and automatically administer treatment. Those suffering from diabetes, for instance, can have their sugar level monitored by a computer and when the level becomes dangerous receive a computerized treatment.

Most of those who use computerized information to organize their actions toward clients regard the computerization of social structures as empowering. Not all clients agree. Many clients of organizations feel degraded when their treatment is structured by computerized information. The computerization of social structures has increased the social distance between members of organizations who deliver services and their clients. For example, in precomputerized hospitals, the health status of patients was determined by the therapist entering the rooms of patients and observing and perhaps speaking to the patient. In computerized hospitals much of the information about the health of patients is accumulated electronically and automatically entered in a computer without benefit of any interaction between patient and therapist. Instructions from high-ranking therapists, such as physicians, to low-ranking therapists, such as physical therapists, are likely to be transmitted via computer printouts.

In many instances, the empowerment of authorities that has accompanied the computerization of complex social structures has magnified asymmetrical relationships. For example, state authorities in charge of collecting taxes can use computers to correlate reported income with taxes received far more effectively than was possible when tax returns were checked manually. In the United States the federal government used computers to match lists of federal employees with lists of welfare recipients to uncover instances of welfare fraud (Weinberg 1990:106). When computers are used in this manner it increases the accountability of subordinates to superordinates.

There has been a series of court cases in the United States that centered on the use of computers by state authorities to monitor the actions of citizens. In most instances the court rulings have attempted to stay the power of state authorities. For example, in 1971 the FBI was ordered to cease providing criminal histories to other agencies (Westin and Baker 1972:60), and in 1974 Congress prohibited the creation of a centralized national information system. Despite the court rulings and legislative enactments, the hand of authorities continues to be strengthened by the computerization of state structures. In 1984 Congress established a de facto national information center (Laudon 1986:5–7) that accumulates and makes available vast quantities of data to state authorities.

Several legislatures have enacted laws that assert the right of citizens to have access to personal information retained by bureaucracies. Congress passed the Freedom of Information Act, which hypothetically gave citizens the right to obtain copies of governmental files about themselves, but many citizens who have sought that information have found it very difficult to fight their way through the bureaucratic maze.

In computerized bureaucracies, lower-echelon clerks "spend the better part of each day with attention fixed on luminous electronic numbers and letters" (Zuboff 1988:124) and interact with one another even less frequently than when they processed information with typewriters and file cabinets. The computerization of bureaucracies has also been accompanied by a decline in the frequency of interaction between clerks and supervisors (p. 150). The lack of contact by clerks with one another and with their supervisors is a source of disquiet for lower-echelon clerks when a bureaucracy is computerized. In the words of one clerk of a recently computerized bureaucracy, "The only reality we have left is when we get to talk to a customer" (p. 131). Clerks hired after a bureaucracy is computerized do not experience that disquiet; they take the established social structure for granted. They are not conscious of the increase in isolation and lessening of autonomy of lower-echelon clerks that accompany the computerization of bureaucracies.

Some fear that the computerization of bureaucracies is ushering in a social structure similar to that described by George Orwell in *1984*. Such may be the case, but the nations leading the way in the computerization of social structures often are in the forefront of protecting citizens from being spied on by authorities. Swedish citizens, for instance, do not seem to be threatened by Sweden's highly computerized bureaucracies.

Although the computerization of bureaucratic structures seems to enhance hierarchies, it may be that the overall impact of computers on the social structures of complex organizations is to erode hierarchies. Hackers and others have initiated the creation of bulletin board networks to enhance the dissemination of many kinds of information, including news about computers and their uses. In addition, computerized bulletin boards provide more details about current events than newspapers and news broadcasts. Several commercial corporations and a few state organizations have instituted computerized bulletin boards that provide information on a number of topics for a nominal fee. Most people with access to computerized bulletin boards use the system to obtain and dispense information on topics of limited interests. However, similar systems that specialize in dispensing information about state affairs, including the actions of state officials, have also been created.

Even when computerization is contextualized by complex hierarchies, it may undermine hierarchical structures. Such seems to occur when the production lines of factories are computerized. On-line operators of computerized production lines are more likely to be conscious of extended sequences than assembly line workers and are more likely to reflect on their actions and how their actions fit in with extended processes. One operator noted, "Dealing with information instead of things is very, well, very intriguing. I am very aware of the need for my mental involvement now. I am always wondering: Where am I at? What is happening now?" (Zuboff 1988:75). In contrast, the traditional assembly line worker is likely to be lost in his or her daydreams. Operators of computerized production lines who adopt a reflective standpoint toward their activity are more likely than daydreaming assembly line workers to make assessments of their supervisors. Conflicts between on-line workers and supervisors concerning how problems might be resolved are more common in automated factories than traditional line factories. Some supervisors of automated factories discourage workers from reflecting about production processes (p. 261).

The computerization of information-processing procedures at least sometimes strengthens the hand of on-line employees of complex organizations in their interactions with their superordinates. For example, parts suppliers and librarians routinely use computerized systems to preserve information and often have greater access to relevant information than do their superordinates. When computerization increases the amount of information available to on-line operators, hierarchical structures tend to be eroded.

Several commercial organizations, especially computer companies, have promoted communication to and among on-line operators through computer networks. These companies have been in the forefront in the development of principles of management by objective. That suggests that when the computerization of social structures increases the information available to subordinates and the flow of information among them, the computerization of social structures is antihierachical.

COMPUTERIZED SCHOLARSHIP

Scholars of past generations validated their identities by displaying books; many modern scholars validate their identities by displaying their information-processing machines. Grammar school students learn to spell with a computer instead of from a spelling book. University students learn everything from how to solve equations to how to design highway systems by interacting with information-processing machines. Time that was spent reading books is now spent using computers.

It is cheaper and more effective to preserve and dispense knowledge via computers than via books. Information can be stored in computers for a fraction of what it costs to put it in books. To acquire information preserved in books, it is necessary to handle the books; to acquire information preserved in computers one need only have access to a computer terminal (which are becoming ubiquitous).

Far more knowledge is now available than when print was the dominant means for preserving knowledge. The amount of information preserved in the British Museum Library and the Library of Congress pales in comparison to what can be preserved by recordings and computers. Computers have also altered the kind of information and knowledge that is preserved and dispensed. For example, when computers are combined with recordings, then dynamic as well as static information can be preserved. A printed dictionary usually offers only words when defining a word, but a computerized dictionary can offer printed words, play audio recordings of the spoken word, and display dynamic configurations.

Information preserved in hypertext can be located more rapidly than information stored in books. If, for example, all the past issues of a scientific journal were preserved in hypertext, then a researcher could go from one article to all other articles referenced in the original article by punching a series of keys while seated at a terminal. In addition, the contents of hypertexts can be modified with far less effort than the contents of a book.

The interactivity with information preserved in computers is greater than the interactivity possible with printed information. When grammar school students use computers to acquire skills in manipulating mathematical concepts, they can be informed immediately whether or not they have used the concepts correctly. Learning to use computers to acquire and manipulate information requires extensive subordination, but once computer formats are mastered they provide greater opportunities than writing and print.

The acquisition and manipulation of information usually is more rapidly paced and more exciting than is the case for written and printed information. For example, computerized instructional programs usually elicit more excitement than books. When students use computers to master a body of knowledge, they often become more excited than when they acquire a body of knowledge from books.

The computerization of data management procedures has opened new vistas for scholars. Yet the computerization of scholarship has been a mixed blessing. Some scholars have become so intrigued with computers that they "generate subproblems of a strictly computational nature" and lose sight of their original objectives (Weizenbaum 1976:36). A few researchers have taken for granted the formats offered by computer specialists and have forced their data into categories offered by computers instead of designing computer formats appropriate for their data. One silly example of that is the Interactional Process Analysis devised by

Robert Bales (1950), which categorizes human activities into twelve categories. The twelve categories were derived from the IBM cards of the time, which had only twelve slots. Is it necessary to say that Bales's research failed to increase our understanding of social interaction? The mistake of equating research with computerized manipulation of quantitative concepts has become less common as techniques for the processing of graphic and narrative information have been perfected and researchers have become more sophisticated in what can and cannot be done with computers.

The distance between researchers and the phenomena under investigation has increased as research has been computerized. For example, instead of examining the flow of fluids through organic tissues, a modern biological scientist is likely to examine simulations of the flow of fluids. In a parallel manner, a sociologist researching social hierarchies and information flow is likely to analyze computerized simulations of information flow instead of observing members of organizations communicating with one another.

Computerized procedures often allow researchers to pose questions not previously possible, but they have also enticed some to employ inappropriate research procedures. For example, the computerization of sociological research in conjunction with the computerization of record-keeping procedures has made bureaucratic records readily available to sociologists. The ease with which such records can be processed with computers has enticed many social scientists to accept the categories developed by bureaucrats as sociological concepts. That procedure presumes that the categorization procedures of bureaucrats are adequate for sociological research, which, as argued in Chapter 6, is a highly questionable presumption. Simply put, when researchers use computers to extend their imaginations and create novel formats for the processing of data, it often facilitates the formulation of more comprehensive understandings of phenomena. But, conversely when computer formats are taken for granted by researchers, the computerization of research can suppress innovative thought.

PRODUCTION

For the past several generations, production was more or less equated with activity that created food, fiber, minerals, or manufactured objects. The production of authors, record keepers, and researchers, however, is not measured by bushels, tons, or the number of objects manufactured. Instead, it is the quantity and quality of information produced, transmitted, or transformed that are used to measure productivity. The development of computers, though, has also transformed the production of food, fiber, and manufactured objects. Dairy farmers record the production rates of their cows and organize much of their activity toward their herds on the basis of computerized information. All but the smallest factories use computerized records to assess and program manufacturing procedures.

The computerization of factories has transformed manufacturing processes. Traditional assembly lines delivered objects in front of workers, and each worker produced a sequence of acts that modified the objects. The sequences usually were of short duration, standardized, and repetitive. The required sequences of action became habitual, requiring no thought. Most assembly workers daydreamed to maintain their sanity. On automated assembly lines, however, computerized machines replace workers and operators monitor terminals instead of objects and manipulate keyboards instead of tools. Referential concepts displayed on a terminal have replaced sensations as the primary source of information. For example, workers in a pulp mill noted that before the factory was automated they listened to the sounds of the boilers, looked at the color of the flames, and noted the smells to acquire information about how the production process was proceeding. After automation they visually monitored how the process was proceeding on a terminal and used that information to make adjustments through a keyboard. When on-line workers begin using terminals, nearly all of them regard the information displayed at their terminals with skepticism, but most quickly learn to rely on it (Zuboff 1988:63).

Many productive processes have moved from automated procedures to robotized ones. Robotization occurs as mechanical devices are developed that obtain information from the immediate environment, relate that information to programs stored in a computer, and act toward select objects. The information obtained by industrial robots may include the size of objects, temperature, and the force required to move an object in a given direction.

The development of robotized production has relieved human beings from some dangers, but several dangerous industries make very little use of robots. For example, robots are not widely used in farming, mining, and construction. But military specialists have developed many robots. Perhaps only robots will be destroyed in future wars.

On a robotized assembly line, the robots respond to the appearance of objects and act on them as programmed and make readings of the consequences of their actions. For example, a painting robot notes the presence of an object to be painted, paints the object, reads the thickness of the paint on the object, relates that reading to a preprogrammed criterion, and depending on the outcome of comparison either adds another layer of paint or waits for the next object to appear.

The industrial segments of industrialized nations are undergoing profound changes as production processes are computerized, but perhaps the changes in the production and management of information that are accompanying the maturation of information-processing machines are even more profound. Industrial production, for the most part, continues to occur in factories, whereas much of the production and management of information has become decentralized.

The distinction between home and place of work has been eroded as more people have acquired personal computers and used them to produce and process information in their homes. The movement of work from the office and factory to the home saves space and overhead for the employer and the cost of commuting for the employee. Personal autonomy also is enhanced. But solidarity among workers is eroded when the workplace migrates from offices and assembly lines to homes.

RECREATION

Computerized video games, which once were an intriguing activity limited to a handful of hackers, have become a fascination for millions of people. In 1982 the amount of money spent on video games surpassed that spent on movies and records in the United States (Turkle 1984:65). Computer games have contributed to the movement of entertainment from a collective context to a solitary one. Others may be present as an audience when computer games are played, but the interaction typically is between one player and a machine.

But in contrast to entertainment derived from recordings, the experiences derived from playing computer games are based on actions as well as responses. Video game players structure their actions by projecting a future and then acting to bring that future to fruition. Video game players organize their actions to achieve a particular objective: to win, or at least score as many points as possible. As video games unfold, players assess the present on the basis of a projected future. The experiences derived from video games include relating the present to the future, they are future-centered. The experiences derived from recordings are more likely to be present-centered.

Action taken while playing a computer game also includes assessments of the adequacy of one's performance and is infused with emotion linked to self. Pride and mortification are generated. Those emotions are not generated when enraptured by recordings. When one is enraptured by a recording, one is emotionally stirred, but the emotions are not reflexive. Players of computer games are detached, analytic, and reflexive; comparatively speaking, consumers of recordings are involved, passionate, and nonreflexive.

Video games are competitive but the activity is more orderly than competition between individuals. When individuals compete with one another, each has the power of initiation. In player/computer competition only the player initiates, the computer only responds. The parameters of responses offered by video games are precise and preset. The parameters of games of competition between individuals are fluid and problematic; those of computer games precise and stable. When players can determine the parameters of video games they frequently become disenchanted with the games.

CONCLUSIONS

Just as the invention of written records transformed the lives of the nonliterate as well as the literate, computers have transformed the lives of those who do not use them as well as those who do. The treatment of hospital patients who have no understanding of computers has been transformed as surely as has the treatment of those who design computers.

Computers are *idiots savants*. When properly programmed they can process some kinds of information more rapidly and accurately than human beings can. They relieve human beings of the drudgery of many boring tasks and provide opportunities for people to participate in new forms of exciting activity. But they also can be used to enhance old forms of constraint. Some critics charge that computers are the ultimate iron cage for humanity.

Computers become iron cages only when they are encased within particular social structures or when users of computers put themselves in cages offered by computers. Computers are cages for those who imitate Pythagoras and proclaim that all is number. Just as Pythagoras attracted a following despite the ridiculousness of his claim, computer users who proclaim that all is number and call for all to subordinate themselves to computer formats also have attracted a following. If they had their way, some computer specialists would place all, including themselves, in iron cages. The adoption of an analytical standpoint toward computers, their use, and consequences of their use can prevent that from happening.

Some in the forefront of computer development visualize a world where electronic sensors monitor everything from global weather patterns to the temperature of a baby's mattress and feed the information into a computer linked to a machine programmed to create an ideal environment. Some think computers are the pathway to Utopia.

Most information technologies were developed to more effectively preserve or transmit information. An unintended consequence of information technologies often has been the altering of the substantive content of the information preserved and transmitted by a given technology. In contrast, computers are designed both to alter information and to process it. The original designers did not intend to use computers to preserve and disseminate information. The preservation and dissemination of information via computers are secondary consequences, but those consequences may have greater significance than the primary uses.

McLuhan argued that as printing replaced orality, the natural ratio of sensory experiences was destroyed, the eye replaced the ear, and human life became less expressive and compassionate. Computerization has been accompanied by an increased emphasis on referential symbols, while recording and broadcasting technologies enhance evocative symbols. Two competing cultures have arisen. The recording and broadcasting culture is evocative, dynamic, and concrete; the culture of computers is referential, enduring, and conceptual.

Orality and literacy overlapped, but literacy eventually supplanted orality. There are no indications that computers will supplant recordings and broadcasting. Indications are that human beings will continue to use both sets of technologies. Perhaps computer technologies will become so merged with recording and broadcasting technologies that the electronic technologies will become unified into an organic whole. Or maybe a frame of thought and action derived from computer technologies will compete with one derived from recording and broadcasting technologies for hegemony and create competing life-styles. Perhaps humanity will become suspended in cyberspace.

14

Knowledge Centers

Like bureaucracies, knowledge centers are second-order information technologies and are encased by larger social structures. Until the twentieth century, with few exceptions, knowledge centers were adjuncts of religious and state structures and their faculties and students were accountable to religious and state authorities. The degree of accountability of faculties and students of contemporary knowledge centers to external authorities is highly varied. In some instances their activities are closely monitored and deviant actions are sanctioned, while in other instances they are subject to minimal surveillance and their actions are seldom sanctioned.

Those who support knowledge centers presume that the knowledge generated, preserved, and transmitted contributes to the welfare of the communities that encase them. All knowledge centers transmit principles of belief and behavior that are presumed to be useful to students after they leave the knowledge centers. For example, the numeric concepts mastered while in school inform the actions of shoppers, carpenters, and astronauts. The actions of physicians and nurses are informed by the knowledge dispensed by medical schools. The knowledge dispensed at knowledge centers impacts on those who do not attend knowledge centers as well as those who do. For example, the knowledge transmitted at medical schools impacts on the clients of physicians and nurses as well as on physicians and nurses.

Each knowledge center contains a relatively stable faculty and transitory students. The faculties and students have an asymmetric relationship based on the presumption that the faculties have command of pools of knowledge that are valued by the students. Faculty/student relationships move toward symmetry as students demonstrate mastery of the knowledge dispensed by the faculty. Faculty/student relationships usually are terminated when students demonstrate command of a specific pool of knowledge or when the faculty make the assessment that a student will not acquire command of a pool of knowledge. For example, high school students are promoted to the next level of mathematics instruction when they demonstrate command of a given set of mathematical concepts. The students judged incapable of mastering the concepts may be dismissed from the knowledge center.

The faculty use information technologies ranging from chalk drawings to computers to facilitate the transmission of knowledge, and much of the knowledge transmitted to students focuses on how to use information technologies to manage knowledge. Students in grammar school learn how to read and write, while those attending professional schools learn many complex procedures for preserving, processing, and transmitting knowledge.

Knowledge centers have an ancient past, but with few exceptions they did not occupy a central position in national and international social structures until the twentieth century. Contemporary authorities and spokespersons of nearly all nations, democratic and totalitarian, advance the claim that viable knowledge centers are critical to national welfare. Nearly all agree that international hegemony is highly contingent on the viability of a nation's knowledge centers. Many nations entice students from other nations to attend their knowledge centers, and, reciprocally, many nations with knowledge centers of limited sophistication send their young people to study in nations with more sophisticated knowledge centers. Over one-half of the students in engineering in the universities of the United States are foreign-born.

THE CORNERSTONES OF KNOWLEDGE CENTERS

Prior to the development of information technologies, there were no knowledge centers. Societies that did not have information technologies did not preserve sufficient amounts of knowledge to justify knowledge centers. Knowledge centers emerged as information technologies were perfected and the amount of knowledge preserved increased. The faculty of the earliest centers used orality, alignments, and markings to preserve and transmit knowledge, while the faculty of contemporary centers use books, recordings, and computers. Regardless of the form, information technologies are the cornerstones of all knowledge centers.

The amount of calendric knowledge preserved had incrementally increased for tens of thousands of years (Marshack 1972) before the first knowledge centers were established to facilitate its accumulation, preservation, and transmission. After ancient knowledge centers were established, the amount of knowledge about celestial phenomena expanded far more rapidly than it had in the preceding millennia. Several ancient knowledge centers created pools of knowledge about many other phenomena in addition to celestial phenomena. For example, the knowledge specialists associated with the monumental centers of ancient Britain subsequently generated the knowledge necessary to construct complex monuments as well as calendric and astronomic knowledge (Thom 1967, 1971). In other instances, the faculty of knowledge centers generated the knowledge necessary for the construction of complex networks of irrigation canals.

Centers of knowledge, called tablet houses, that preserved and dispensed knowledge of record-keeping procedures emerged in Mesopotamia late in the fourth millennium B.C. The centers of knowledge dominated by timekeepers and those dominated by record keepers coexisted independently of one another for at least a few hundred years before they were merged. The first knowledge center that merged calendric and record-keeping knowledge probably was Heliopolis, which was established in Egypt about 2700 B.C.

After written languages were developed, scrolls, followed by books, became the dominant artifacts of knowledge centers. Knowledge centers mushroomed in Europe after the invention of print. There has been an equally dramatic mushrooming of knowledge centers on a worldwide basis in the twentieth century as recordings and computers began to compete with books for hegemony in the preservation of knowledge.

Although knowledge centers rest on a foundation of information technologies, the faculty of knowledge centers seldom have been in the forefront in the development of information technologies. Techniques of time factoring were developed long before knowledge centers were established (Marshack 1972), and numeric concepts and written languages were developed by bureaucrats, not specialists affiliated with knowledge centers. More recently, the faculty of knowledge centers have contributed very little to the invention of print, telecommunications, recordings, or broadcasting.

More often than not, it has been only after a technology has achieved a fairly widespread acceptance that the faculty of knowledge centers have become active in refining a technology. For example, recording and broadcasting technologies achieved almost universal acceptance before the faculty of knowledge centers began using them to preserve and transmit knowledge and instructing students in their use. Many of the faculty of universities continue to look askance at departments of broadcasting and motion picture production. Knowledge centers with extended histories have usually been more resistant to teaching new technologies than newly established ones.

Asymmetric relationships wherein the more knowledgeable instruct the less knowledgeable have prevailed in knowledge centers from their beginnings. Some dimensions of the teacher/student relationship have varied across time and space, but throughout millennia students have subordinated themselves to teachers and teachers have offered definitions of reality for students. For example, five thousand years ago the teachers and students of Sumer related to one another in much the same way as do teachers and students of modern knowledge centers (Kramer 1959:8–11). The techniques used to transmit information, however, have undergone major changes. For example, the spoken word is somewhat less dominant at knowledge centers than it was hundreds of years ago, and memorization no longer is stressed as much as it once was. But as has been the case for thousands of years, it is the faculty who offers definitions of reality and principles of conduct that students are expected to learn and use.

KNOWLEDGE CENTERS, TEMPLES, AND STATES

Many ancient nonliterate civilizations contained temple-centered social structures that rested on a foundation of knowledge generated by observations and analyses of celestial phenomena and were preserved in the alignments of temples and surrounding monuments (Couch 1984:155–70). The ancient Britons were but one of many societies whose social structure was keyed to monuments and temples (Thom 1967, 1971; Thom and Thom 1978; MacKie 1977a, 1977b). The faculty of these centers usually dispensed knowledge in a carefully orchestrated manner, and the knowledge dispensed elevated the faculty from the rank and file members of the communities. For example, the faculty of the knowledge centers of ancient Briton were relieved from manual labor, had a diet superior to that of other members of the community, and were given special burials (MacKie 1977a:151–76). The rituals conducted by these specialists not only informed the community that the faculty were extraordinary people, they also provided a temporal order for the community and affirmed communal unity.

The faculty of some, perhaps nearly all, of the ancient nonliterate knowledge centers imbued the knowledge that they preserved with sacredness and only a select few were granted access to the knowledge. As written languages replaced alignments and orality as the dominant procedure for the preservation of pools of knowledge, scrolls and books were imbued with sacredness. For example, in Chinese schools at the beginning of each school day the students bowed toward the tablet of Confucius (Rawski 1979:55). The development of print diluted, but did not entirely remove, sacredness from books. Most Christians find it difficult to spit on a Bible.

Although some knowledge centers endured for thousands of years, others flourished only for a generation or two. In at least a few instances the supporting community became disenchanted with the centers and destroyed them. For example, the ancient Olmec mutilated and buried the statues that graced their knowledge centers. Perhaps they also killed their knowledge specialists. In China in the 1960s and 1970s the Red Guards attacked the faculty of knowledge centers and forced them to do manual labor. There was widespread communal support for the actions of the Red Guards. Many knowledge centers have faded into oblivion when communities have discontinued their support of the centers. A few centers have been destroyed by environmental catastrophes and many centers have been destroyed by invaders.

From their beginnings until a few centuries ago, most knowledge centers were adjuncts of temples. Such was the case for such diverse civilizations as the Sumerians, the Maya, and the Europeans of the Middle Ages. One exception seems to have been Heliopolis, an Egyptian knowledge center established as an adjunct of the state in the Second Dynasty (about 2700 B.C.). Knowledge centers as adjuncts of state structures did not become common until recently.

Within a century or two after Heliopolis was established, its faculty perfected a written language, formulated three-dimensional analytic geometry (Badawy 1965, 1966; Reese and Couch 1990), developed geometric sculpture and relief carving, invented the first numeric calendar (Parker 1950, 1974), and developed anatomy and medicine. The records from the Third and Fourth Dynasties refer to Heliopolis almost as often as they do to the national capital (Kees 1961:147), which suggests that Heliopolis occupied a central position in the social structure of ancient Egypt. The inquiring spirit that prevailed at Heliopolis was not matched until the Golden Age of the Greeks of the fifth century before the Christian era. The innovative activity at Heliopolis drastically slowed when Heliopolis became the national capital at the beginning of the Fifth Dynasty and the state structure of Egypt became highly bureaucratized (Baer 1956).

When other ancient civilizations established knowledge centers as adjuncts of state structures, there was, without exception, an explosive burst in the knowledge generated, but within a generation or two innovative activity stagnated. On several occasions the state authorities of China established knowledge centers with the centers becoming the source of innovations for a few decades followed by bureaucratization and stagnation (Nakayama 1973:34). It seems that the first one or two generations of faculty of the knowledge centers that were adjuncts of state structures were intellectuals first and bureaucrats second; the later generations were bureaucrats first and intellectuals second.

When state authorities establish knowledge centers as adjuncts of state structures, they anticipate that the knowledge generated will enhance the prestige of the state elite. For example, the geometric concepts and engineering principles formulated by the faculty at Heliopolis were used to construct the great pyramids that enhanced the prestige of both the state elite and the knowledge specialists. In a similar vein, the interests of modern American scientists and engineers merged with the interests of American state authorities when a program for the exploration of outer space was undertaken.

Despite the close interweave of knowledge centers with religious and state structures of ancient civilizations, the relationships between knowledge specialists on the one hand and religious and government authorities on the other hand often were antagonistic and conflictual. Religious and state authorities on several occasions have instituted the wholesale slaughter of suspected heretics among knowledge specialists. Reciprocally, knowledge specialists have often criticized state and religious authorities. It may be that in the long run the pen is mightier than the sword, but in the short run the sword has often prevailed over the pen.

THE GREEK ACHIEVEMENT

The uniqueness of the knowledge generated by the Greeks during the Golden Age of the fifth century before the Christian era has been exaggerated, but the distinctiveness of the social context in which the knowledge was generated has been underappreciated. The intellectual turbulence of the Golden Age of the Greeks was achieved without benefit of knowledge centers. Neither Plato, whose writings mark the culmination of the Golden Age, nor his precursors were educated at knowledge centers.

Instead, most of the intellectual activity of the Golden Age of Greece occurred at the markets where the Sophists sold knowledge much as grain dealers sold grain. "There was no such thing as a standard sophistic curriculum of studies" (Kerferd 1981:37) nor were there schools that taught rhetoric, literacy, the arts, or the sciences. Instead, Sophists offered knowledge to clusters of clients for a fee. The Sophists met with their clients in private homes, public places, and secluded gathering places. The term *Sofia* originally referred to the wisdom offered by nonliterate sages (Kerferd 1981:124), but during the fifth century before the Christian era the term came to be applied to those who offered knowledge for a fee. The practice of the Sophists of selling their knowledge to all kinds of people enticed some to condemn them (p. 25), but the term did not have derogatory implications until after Plato. It was the Sophists, independent mercenary scholars, more than any other category of Greeks, who generated and dispensed knowledge during the Golden Age of Greece, although most of them barely eked out a living by selling their knowledge to clients.

The museum of Alexandria, created about 330 B.C., was the first major Greek knowledge center supported by the state. The relationships that prevailed between faculty and students at the museum were similar to those that had prevailed in other ancient knowledge centers. The Sophists referred to the faculty of the museum as fowl in a chicken coop. The museum of Alexandria became rife with corruption within a generation after it was established, but it endured as the leading knowledge center of the region until it was destroyed by fire after the Romans conquered Egypt.

Some of the knowledge dispensed by the Sophists diffused into other societies. Arabic knowledge specialists preserved some of the knowledge that the Egyptians and Greeks had generated, and when the Muslims invaded Europe some of that knowledge diffused to European scholars. That knowledge and the knowledge that had been generated and preserved at European centers were the kernels that flowered into universities during the Middle Ages.

THE UNIVERSITY AS A SOCIAL FORM

The first universities were created about 1200 A.D. They were preceded by *stadium generals,* which were irregular clusters of scholars that had emerged in several commercial centers in the preceding centuries. The scholars of the stadium generals formed universities for "very much the same reasons as merchants and artisans formed universities—to protect their interests, to bury their members when they died, to look after their widows" (Hyde 1988:18). The medieval term *university* referred to "a group of men engaged in a common activity of any sort" (Ferruolo 1988:24). "It is a mere accident that the term has gradually come to be restricted to a particular kind of guild" (Rashdall 1936:5).

The charters of the early universities gave prerogatives to foreign scholars that were denied other foreign residents of a city. For example, the charter of the University of Bologna gave foreign faculty and students the same legal safeguards as held by citizens of Bologna (Hyde 1988:18). As universities became more common, scholars "were frequently exempt from taxes, levies, tolls, and military service" and students often were accountable only to their faculty (Baldwin 1971:46).

Conflicts between scholars and other residents of university cities, at least in principle, were to be resolved by spokespersons from the university negotiating with civic authorities. For example, the first recorded

gown/town brawl in Europe occurred in 1192 in Paris, when "one scholar was killed and several others injured by serfs of the nearby abbey"
(Ferruolo 1988:31). The university faculty called on the civic authorities to
intervene and the charter of the university was renegotiated. The new
charter specified that all citizens were obliged to report crimes by laymen
against scholars, that scholars would be arrested for only the most hei-
nous crimes, and that no force could be used when arresting scholars (ibid.).

Universities commonly threatened to relocate when conflicts emerged
between local citizens and students, and on several occasions universities
relocated when they were not able to negotiate the arrangement with
local authorities that they desired. "There were major migrations of scholars from Oxford to Cambridge (1209), and from Bologna to Vicenza
(1204), to Arezzo (1215), and to Padua (1222), and then from Padua to
Vercelli (1228)" (p. 23).

The charters negotiated between the communities of scholars and city
officials presumed a symmetrical relationship between two corporate
units, namely, the host city and the university. The threat and practice of
relocating combined with the practice of the faculties of universities playing the sacred and secular authorities off against one another contributed
to the universities obtaining a degree of autonomy from state and religious authorities and the surrounding communities.

The early universities were open to nonbelievers as well as believers
(Hyde 1988:13), whereas all the scholars affiliated with monasteries pro-
fessed to be believers, which created a distinctive context for learning. For
example, the students of the University of Bologna studied both canon
and civil law and argued about their respective merits. The superiority of
canon law was taken for granted in any presentation of civic and canon
law in a monastery. The universities created "semicloistered heterogeneity in the midst of uncloistered heterogeneity" (Bender 1988b:290).

The universities of the Middle Ages were both a source of community
pride and irritation. Universities contributed to the prestige and wealth of
the host communities, but many citizens resented the special prerogatives
extended to scholars and the universities were often the source of considerable turbulence. Student riots were common and local residents "complained that the safety of their women and the peace of their houses were
continually threatened by marauding student bands" (Baldwin 1971:49).
The residents of many contemporary university towns continue to look
upon the presence of university students with ambivalence.

During the reformation, many universities were the targets of religious
authorities who sought out heretics. The Protestants were as diligent as
the Catholics in their search for heretics. In sixteenth century, all the
professors of Leipzig who would not accept the teachings of Luther were

dismissed (Rudy 1984:68). The intellectual viability of many universities declined as authorities sought out heretics.

Closely interwoven with the religious conflicts were power struggles between urban and national interests as nation-states emerged in Europe. Often the universities became pawns in complex struggles between state and urban interests and between religious groups. On occasion the autonomy of universities improved as the consequence of the power struggles, but more often it was depleted. The universities of France were almost destroyed when Napoleon made them adjuncts of the state. Ever since the emergence of nation states in Europe, it has been a common practice for monarchs and dictators to make universities adjuncts of the state. Nearly always when that has occurred universities have been sapped of their intellectual vitality (Bender 1988b:293). Despite the many attacks suffered by universities, however, they have endured although often as stunted intellectual centers. The belief that universities should be marketplaces of ideas became viable in the intellectual circles of Western Europe. The concept of academic freedom was first formulated in Holland after a lay board of regents was established to supervise Dutch universities.

The efforts of dictators and monarchs to make universities adjuncts of state structures has been partially countered by advocates of legislature-centered state structures calling for states to create universities that have a degree of autonomy from religious and state authorities. For example, after the American Revolution, some of the leaders of the revolution agitated for the establishment of a national university. Washington left a bequest to aid in the establishment of a national university. Jefferson established the University of Virginia in 1825, which set the mold for the state-supported universities of the United States. In principle, the faculty and students of the University of Virginia were not subject to the control of state authorities, and, as it turned out, the students there were as undisciplined as their European counterparts of earlier generations. For example, in 1839, a University of Virginia professor was horsewhipped by students.

During the nineteenth century many Western religious leaders initiated the creation of church-supported universities to combat the secularization of knowledge. The contest for control of universities swung in favor of secular interests in the United States when land grant universities were instituted in 1862. The land grant universities were charged with the obligation of transmitting knowledge on how to care for plants and animals to farmers and their children. American farmers subsequently surpassed those of all other nations in their effectiveness in producing food and fiber.

In the latter half of the nineteenth century, Japanese universities were transformed by a governmental edict that required all universities to teach the practical sciences. In a similar manner, after the Communist revolution, the USSR established knowledge centers that specialized in transmitting practical knowledge. In these and similar instances, there was a substantial increase in the number of people who attended centers of higher learning, but very few of them became sources of innovation. With the exception of those established in North America and parts of Western Europe, most universities established in the nineteenth and twentieth century were conceptualized as adjuncts of the state and closely monitored by state authorities.

After World War II, several nations established universities modeled after the land grant universities of the United States. The U.S. government and the governments of a few other wealthy nations facilitated the establishment of universities in the less wealthy nations. Nearly all authorities of nations with well-developed knowledge centers regarded the establishment of knowledge centers that would diffuse knowledge to the less wealthy nations as an effective strategy for extending their influence.

The knowledge centers that have been established in the twentieth century in developing nations have facilitated the diffusion of knowledge across national boundaries and greatly increased the number of people attending knowledge centers. But it is questionable that these new knowledge centers have enhanced the influence of the wealthy nations. In many instances national movements to lessen the control of wealthy nations over poor nations have be fueled by ideas stemming from the newly established knowledge centers. The British Empire might still be viable had the British not been so diligent in educating their foreign subjects. In many instances revolutionists educated at these knowledge centers organized successful rebellions that put in place state structures populated by natives. The authorities of these newly created state structures often made the universities adjuncts of the state structure with little or no autonomy.

AUTONOMY

Until the emergence of universities in Europe, with few exceptions, it was taken for granted that the faculty and students of knowledge centers were accountable to external authorities. That condition prevails today in most regions of the world for grammar schools, high schools, religious colleges and universities, and most state-supported universities. But as the number of universities modeled after the European ones has increased, there has been an accompanying increase in the number of people advocating a low degree of accountability of faculty and students of universities to external authorities. The degree of autonomy that universities should have has become a worldwide public issue. The issue of the degree of accountability is for the most part limited to centers of knowledge that generate knowledge as well as preserve and transmit it. Very few regard a low level of accountability to external authorities as necessary or even desirable for centers of knowledge that dedicate the bulk of their effort to the preservation and transmission of knowledge.

Although the issue of autonomy has had some currency in Europe since the thirteenth century, it has only been a viable issue in very limited circles until very recently. Throughout the first half of the twentieth century when state authorities increased their control of universities, very few questioned the actions of state authorities. Most citizens have been indifferent to the welfare of universities and among those familiar with universities many are of the opinion that university faculties are slothful, the source of heresies, and do not merit greater autonomy than other citizens.

Nor has the faculty of universities always been resistant to state control of faculty and students. For example, when the Nazis assumed control of the German universities there was no organized resistance by the faculty or students (Gallin 1986). In contrast, when the Nazis assumed control of the journalism and labor unions, there was organized resistance by some union members. In a similar vein, only a very small percentage of the Chinese faculty resisted the attacks on university faculty in the 1960s and 1970s. Nearly all of the faculty of Soviet universities accepted state control of their activities without protest (Shlapentokh 1990). On occasion, a few of the faculty have resisted increases in control of universities by external authorities, but the dismissal of one or two resistant members has nearly always brought the rest of the faculty to their knees. Only rarely have state authorities had to resort to violence to elicit compliance to state authorities by university faculty.

Although most university faculty have complied with the dictates of state authorities, the paranoid attitudes of dictators and monarchs toward universities has been well placed. The faculty of universities frequently generate and dispense definitions of reality that contradict traditional definitions of reality. When conflicting definitions of reality have currency within a society, they often are the source of social turbulence. That is especially likely to occur when the state structure is a dictatorship or monarchy justified by traditional definitions of reality. However, when conflicting definitions of reality have currency, social turbulence does not automatically follow. For example, in Germany while the majority of the faculty of Germany universities regarded the characterization of how race and culture are intertwined offered by the Nazis as bizarre, almost none publicly indicated their disagreement with those characterizations. A few faculty regarded the Nazi views as valid (Gallin 1986). In a similar vein, when Lysenko's biological theory had currency, the Soviet authorities decreed that Lysenko's definition of reality was the valid one. Biologists on the faculty of the universities of the Soviet Union who thought otherwise were a silent majority for several decades.

There has been a tremendous increase in the amount of knowledge generated at universities in the past few decades. That growth stems, in large part, from increases in the autonomy and funds allocated to them. Most of the increase in funds allocated to universities is the consequence of states investing more resources in universities than ever before. Those increases have been a mixed blessing, however, since they have been accompanied by increases in the application of bureaucratic criteria to intellectual activity. When the faculty of universities compete with one another for state funds, the parameters of the competition are specified by bureaucratic criteria. The subordination of intellectual activity to bureaucratic criteria may result in a stagnation of the advanced centers of learning similar to that experienced at Heliopolis when the Fifth Dynasty was installed in ancient Egypt.

The vitality of universities for the past several centuries has rested on their heterogeneity. The heterogeneity of universities has been eroded by the emergence of special research institutes that focus on narrow sets of issues. For example, the growth of the research centers associated with the computer industry in the Silicon Valley was dependent on symbiotic relationships of the centers with universities, but as the research institutes have achieved success the heterogeneity of universities has lessened.

HEGEMONY OF KNOWLEDGE CENTERS

The belief that human welfare is contingent on the viability of knowledge centers is so widespread that knowledge centers compete with state structures for hegemony in specifying the parameters of human life. Even the Chinese Communists, who routinely denounce foreign influence, accept the fact that the future development of China is contingent on sending Chinese students abroad to be educated at foreign knowledge centers (Hayhoe 1989:56).

Knowledge centers have neither military might nor economic wealth. Their hegemony stems from the ability of the faculty to formulate and transmit definitions of reality more effectively than the elite of other social structures. A large percentage of the world's population seeks out the knowledge dispensed at advanced centers of knowledge and there is a worldwide movement to make attendance at grammar school universal. The desirability of universal education acquired some currency in intellectual circles during the sixteenth century in Europe, and in the seventeenth century, the Chinese authorities issued an edict calling for all communities to establish primary schools (Rawski 1979). Movement toward universal eduction began in the United States during the revolutionary war. "In 1789 the Massachusetts legislature required every town with at least fifty families to have schoolmasters" (Cmiel 1990:47). Universal access to elementary education was achieved in a few nations in the nineteenth century, but it was not until the twentieth century that universal education became a common national goal.

A substantial minority, in many instances approximately one-fourth, of a nation's citizens attend advance centers of knowledge. In addition, the flow of knowledge from books to recordings and computers is making knowledge that was once reserved to a limited few available to multitudes via telecommunications. Other developments that are enhancing the hegemony of knowledge centers include the closer integration of advanced centers of knowledge with grammar and high schools, the linkage of the faculty of various universities with one another by computerized telecommunication networks, and the formation of cyberspace communities centered on common interests.

Knowledge centers have surpassed the temple-centered social structures from which they originally emerged in their impact on human life. Elite specialists in knowledge have a far greater impact on structuring the experiences and actions of humanity than do religious elites. What once was an adjunct of another type of social structure has become a social structure in its own right that towers over its parent.

CONCLUSIONS

Knowledge centers have traveled a rocky road, but have become social structures that compete with state and economic structures for hegemony in programming the future endeavors of humanity. The program to explore outer space was formulated in a series of complex negotiations between delegates from the state and knowledge centers. National and international programs dedicated to enhancing the health of humanity are formulated in a similar manner. The elite of knowledge centers have surpassed the elite of temple-centered social structures in their impact on programming the future.

If all the knowledge preserved at knowledge centers were to fade from human consciousness, humanity might not survive. Knowledge centers have become the linchpin for current social structures. Contemporary state, market, and bureaucratic structures are more dependent on knowledge centers for the trained personnel to maintain themselves than ever before. The continued vitality of knowledge centers, however, is not assured. The elite members of knowledge centers seek to ensure their selfish interests as surely as do the elites of temples, palaces, bureaucracies, and legislatures, and many knowledge specialists refuse to acknowledge their symbiotic relationship with those who support them. A failure to acknowledge their dependence on and debt to the community that encases them may be their undoing.

Most contemporary knowledge centers are adjuncts of state structures, which is both their strength and weakness. They are the recipients of more resources than ever before, but access to those funds is contingent on meeting the criteria specified by state bureaucrats.

Despite many exceptions, for the past few centuries centers of knowledge, especially universities, have nurtured humanism, science, and social responsibility. Perhaps they are laying the foundation for novel social structures that will be more satisfying than tradition ones.

CHAPTER

15

Oh, What Webs Those Phantoms Spin

Whereas several millennia ago humanity consisted of tens of thousands of people living in small nomadic bands, today humanity consists of billions of people linked together by Earth-spanning social structures. If the information necessary to use information technologies were to suddenly fade from our memories, these structures would collapse and most of humanity would die within a few months. The survivors would exist in small clusters in threat of extinction. In light of that possible consequence, why is it that social scientists have devoted so little effort to the analysis of information technologies and social orders?

Part of the answer is that most traditional social scientists presume that information technologies are phantoms, things apparent to the senses but of no consequence. One cannot help but notice the presence of a book, a television set, or a computer; nor can one fail to notice that most people spend a substantial part of their lives attending to and acting toward information technologies and using the information obtained to organize their activities.

Although most social scientists conceptualize information technologies as phantoms, they treat the content that those technologies preserve and dispense as if they have consequence. Tens of thousands of research endeavors have examined the correlations between what is printed or broadcast with a variety of variables. In contrast, only a few social scientists have focused on how the use of information technologies has modified how human beings relate to one another and to their environment.

The failure of social scientists to recognize that information technologies are consequential in combination with the maturation of those technologies themselves has created a void in the academy. That void has been partially filled by the emergence of departments of communication and media theorists who focused their analytic powers on the electronic media. Many of those specializing in the analysis of communication, though, have followed the lead of their social scientist cousins, and compute correlations between the content of the media and other psychological and social variables. Others, especially the media theorists, attempt to specify how each information technology favors or disfavors the retention and circulation of particular types of information. In the course of all of that, university students flock to departments of communication while departments of sociology suffer from a paucity of students.

A considerable proportion of the analytic power of the media theorists is dedicated to demonstrating that one consequence of the maturation of the electronic information technologies has been a dilution in the value of information. Just as Socrates argued that the information preserved in writing was inferior to that retained orally, most media theorists argue the information preserved and dispensed by the electronic media is inferior to that offered in print.

In the meantime, billions listen to and view broadcasts, millions flock to retail stores to purchase recordings, and book and newspaper publishing is in the doldrums. Much of the world's population, in other words, has been rushing into the information age.

Something is amiss. Either social scientists and media scholars are failing to ask significant questions about information technologies, or humanity, like lemmings rushing into the sea, is on a course of happy self-destruction. Humanity may be on a course of self-destruction and our information technologies may be blazing the path to self destruction for us, but I strongly suspect quite the opposite is the case, namely, that the maturation of the electronic information technologies is opening new opportunities for humanity. New opportunities usually are accompanied by new risks. The development of radio broadcasting provided millions with the opportunity to enjoy pleasant sounds from afar. It also provided Goebbels and Hitler the opportunity to create a state structure infused with a charismatic relationship focused on the head of the German state. Millions died as a consequence of that opportunity.

Some look to social scientists and media scholars for informed assessments of the opportunities and risks that accompany the development of information technologies. As yet, though, the assessments offered have had little if any more merit than those offered by the nonspecialists. If social scientists and media scholars are going to offer assessments that have greater merit than those offered by interested bystanders, they will have to reframe and refocus their analytic endeavors. In particular, it will be necessary for them to do more than compute correlations between the information processed by the technologies and psychological variables and to offer more than pontifications about whether an information technology favors or disfavors a particular mode of thought without benefit of data.

At the minimum, we will have to learn how to observe and analyze people using the media to share information, to form and maintain relationships with one another, to relate to our environment, to preserve information about the past, and to project futures. The accomplishment of that objective requires that we observe and analyze information and information technologies, people using information technologies to preserve and/or share information, and the intertwining forms of social relationships with the formatting properties of information technologies.

The effective use of information technologies can facilitate the accomplishment of greater understandings of the intertwinings of information technologies and other facets of human life. For centuries the written word has been the lingua intelligentsia, and knowledge specialists, with few exceptions, have been slow to investigate the potential of other information technologies for the generation and sharing of knowledge. Many fail to recognize that written languages are but a tool for generating, preserving, and dispensing information and knowledge. Some of the more recently developed technologies though are more effective tools than written languages. If social scientists continue to be reluctant to use the newly developed technologies, it is likely that the social sciences will be pushed off center stage in academia.

QUALITIES OF INFORMATION

Symbolic information has qualities not shared with inorganic and organic phenomena. Consequently, the research methodologies developed to research inorganic and organic phenomena are not adequate for researching symbolic phenomena. The distinctive features of symbolic information include its ephemerality, its ability to rapidly expand and contract, and its ability to be transmitted without depletion of the source.

Matter and energy can be transformed, but can be neither created nor destroyed. In contrast, information is constantly created, destroyed, and lost. The genetic information that informed the reproductive processes of dinosaurs no longer exits. When compared to matter and energy, genetic information is ephemeral. However, when genetic information is compared to symbolic information, it is exceedingly constant. Symbolic information is indeed ephemeral. Only traces of the information that structured the building of the Great Pyramid remain today, while the information that informed that activity has been lost.

The ephemerality of information preserved without benefit of an information technology is so great that it is difficult to subject such information to analytic reflection. Consequently, studies of information, its flow among people, and its flow across generations seldom focus on information that is preserved and transmitted without the benefit of an information technology. For example, when studies of the transmission of information in a face-to-face context are undertaken, researchers usually either attempt to observe the processes without using an information technology to aid their observations, in which case the reliability of their observations may be suspect, or to record with paper and pencil or electronic recording devices the information transmitted. When the latter research procedure is used, one of the central qualities of the information, its ephemerality, no longer prevails.

When analyzed from the standpoint of modernity, with few exceptions increases in the quantity of symbolic information preserved and disseminated have been incremental. However, on occasion there have been explosive expansions in the amount of symbolic information in circulation, and on occasions those expansions have been accompanied by significant transformations in the quality of information in circulation. Conversely, catastrophic devolutions of the amount and quality of symbolic information have occurred and devolutions presumably remain a possibility.

Whereas the transmission of energy and matter diminishes the source, the transmission of information does not. For example, the movement of salt from salt water transforms what remains, but the transmission of information from teacher to student does not transform the teacher. The teacher, that is, remains as informed as he or she was before sharing the information with the student. The sharing of information transforms the relationship of the teacher and student but not the teacher per se. Consequently, research on the flow of information needs to be focused on qualities of relationships, not qualities of individuals.

In a similar manner, when information seeps through boundaries the seepage neither lessens the amount of information at the source nor is the new host depleted. In contrast, when inorganic matter seeps, it depletes or transforms the source and often depletes the new host. For example, when oil seeps to the surface it both lessens the source and depletes its new environment. When information flows across boundaries, though, the configurations of the boundaries and relationships are often transformed. For example, when information about the activities encompassed by a social structure seeps, both the relationships between the original structure and other structures may change as well as the internal structure of the source of the information. Again, research that will increase our understanding of information technologies and human life needs to focus on relationships between social units, not on the qualities of individuals.

To pursue these matters further, it can be said that the belief that the individual is sociology's atom, as Weber argued, is as false as the belief that the Earth is the center of the universe. Dyads and their environments are sociology's atom and the lower limit of meaningful action. Symbols, minds, selves, and social bonds are produced and carried by pluralities, not by individuals (Mead 1934; Hintz 1975). As minds, selves, and social bonds are derivatives of collective acts, the smallest basic units of analysis for students of information technologies accordingly are dyads employing an information technology to enhance the retention and dissemination of information. Although nearly all students of social relationships agree that social relationships are created, affirmed, modified, and destroyed through social interaction, as yet there have been few studies of those processes (Seckman and Couch 1989).

INTERFACINGS OF INFORMATION AND SOCIAL RELATIONSHIPS

Human beings are capable of constructing a vast range of social relationships without benefit of any complex and artifactual information technology (Simmel 1950; Weiland 1975; Couch 1989). A basic dimension of many relationships is shared consciousness of a shared past. Early in life young children become capable of distinguishing strangers from acquaintances. The information that allows them to make that distinction is acquired through copresent interaction. Once the consciousness of the familiarity with another has emerged, nearly all subsequent encounters are contextualized by mutual consciousness of the presence or absence of a shared past and interactional history.

Once a shared past is in place, a particular form of sociation is activated each time those with a shared past establish copresence and mutual recognition. When a tyrant (master) and victim (slave) establish copresence and recognize each other, a complex set of elements of sociation are activated that contextualizes their interaction. Similarly when friends convene, a particular cluster of elements of sociation is activated when mutual recognition is achieved.

The consequences of each communicative act are a function of the social relationships that contextualize them as well as of the acts themselves. Students of communicative processes must maintain a perceptual scope of sufficient temporal depth to include the social relationships(s) that contextualize communicative acts. Just as each form of a social relationship has distinctive properties, so does each medium of communication have distinctive formatting properties that give meaningful transactions a distinctive framing (see Altheide 1985).

THE SEATING OF MEDIA USE

Each information technology discussed in this book emerged within a distinctive context. In some instances, as with Chinese printing, the technology was rather tightly encased within a particular social structure. In other instances, such as emergence of phonetic writing among the Greeks, the emergence of information technologies was not tightly contextualized by particular social structures. Such was also the case for the emergence of books in Europe and broadcasting in the United States. When the maturation of an information technology has not been tightly encased within an extant social structure usually there has been a fair amount of turbulence in its early years. On a few occasions, the maturation of a technology has been partially encased by an extant structure while other facets of its maturation have occurred outside an extant structure. For example, many aspects of the development of computer technologies were clearly under the control of established elites, while other aspects such as the elaborations offered by the hackers occurred outside that control.

Recording technologies have never been effectively encased by any extant social structure, nor have they been subjected to much control from a literate elite. When information technologies have not been effectively controlled by established authorities and elites, they have usually been the source of a fair amount of social turbulence in the early decades of their use. But they usually are subsequently subordinated to a new social structure which may have arisen in part to control them.

Whether information technologies emerge within extant structures or external to them, taken-for-granted linkages eventually develop between them and social structures that have been established. For example, taken-for-granted linkages now prevail worldwide between newspapers and legislative-centered state structures. However, when written language first appeared, those few people who noticed it regarded it as a curiosity that provided the opportunity to render speech into inscriptions. In time it obviously has become a taken-for-granted facet of bureaucratic structures.

When a new technology modifies how symbols are formatted, it is disruptive of the existing configurations and social habits. Social habits associated with information technologies, such as writing and reading, allow people to associate with one another with little effort or reflective thought. Each social habit is both a resource and a constraint since, for instance, the time required to acquire mastery of reading/writing could be spent pursuing other activities. In addition, once mastered, a technology often inhibits the development of alternative social habits, as habits linked to information technologies tend to endure long after other far more efficient ones for preserving and disseminating information have been developed. Pictographic writing, as a case in point, is still viable in China.

Social habits rest on nonreflective memories. That is, memories are activated without reflective thought whenever specific conditions present themselves. Prior to the development of hand-held computers, many merchants added up retail sales manually. Contemporary sales people are more likely to use their hand-held computer, but if it fails then it is necessary for contemporary merchants to add up the sales manually. In such situations of technology failure, it becomes evident that some people lost the ability to think when they became adept in using and dependent upon the new technology. Nonreflective memories thus structure our experiences and actions without our being aware of the structure they provide.

Each time a habitual procedure is rendered problematic, it disturbs the social order. The adoption of some procedures are so disruptive that new technologies are rejected. Although most computer scientists find the base-2 numeric system more efficient than the base-10 system, it is doubtful that the base-2 system will replace the base-10 in the near future. Those who have mastered the traditional technology are so reluctant to devote the time and energy necessary to master a new technology that an old habit continues to endure.

Some claim that the new technologies have created a new tower of Babel; that they have created *No Sense of Place*. Social boundaries were poorly defined in Babel and some media theorists (Meyrowitz 1985) advance the claim that the maturation of the electronic media and their convergence into a single medium have disrupted traditional social orders without replacing them with alternative social orders. However, what has occurred is that now one's location in the media environment is far more critical than one's geographical location.

HEGEMONY OF RELATIONSHIPS

The social structures of societies with few or no information technologies consisted of relatively simple sets of intertwined social relationships. In some of the simple nomadic societies nearly all social activity was contextualized by either solidary or parental relationships. Social structures became more complex as people created religious, bureaucratic, market, and military structures that partially replaced solidary and parental relationships.

Since the emergence of state structures, international hegemony has been primarily a function of the viability of economic, governmental, and religious structures. Some of the contests between those social structures are coming to an end and others are being transformed as novel social structures have entered the contest for hegemony. Social structures emanating from entertainment and knowledge centers compete with the traditional structures for dominance, and market-centered structures have superseded nation-states as the largest social structures. Many of them encase the large bureaucracies and knowledge centers, and the most rapidly growing economic structures are those that produce and process information and knowledge.

The development of electronic technologies has moved the contests for hegemony from the national to the international arena. New forms of imperialism have emerged. Not only do nations compete with one another in the international arena, but Hollywood and American universities compete with national capitals and Wall Street .

Whereas hegemony based on traditional structures is often resented, those based on entertainment and education are usually readily accepted. The hegemony based on entertainment is achieved by offering a product that has a wider appeal than that offered by others. Hegemony acquired through entertainment is based on attracting others. The hegemony of knowledge centers is based on providing access to the knowledge and instruments of knowledge management. The uninformed turn to the informed to acquire repertoires of action. Although the acquisition of entertainment and education require subordination, it is often a willful subordination. When the subordination is willful it is not necessary for subordinates to monitor subordinates in order to maintain a relationship.

In the past, the collapse of state structures has been associated with a decline in food, clothing, and dwellings. For example, when the Maya state structures collapsed, the people became more vulnerable to invasions and natural catastrophes and the quality of life declined. A decline in quality of life, however, does not seem to be accompanying the contemporary decline of state structures. Neither the distribution of goods nor entertainment nor knowledge necessarily entails asymmetries. But when there are a limited number of outlets for a good, entertainment, or education, an asymmetric relationship prevails. The distribution of goods, entertainment, or knowledge can be contextualized by a social hierarchy, especially when the source of the goods, entertainment, or knowledge is capable of creating a monopoly. Most monopolies are created to institute hierarchies. When social hierarchies rest on monopolies of knowledge, much of the asymmetric influence of the elite stems from its ability to program the future.

The technology is available for all to share in the information necessary for programming the future, but the technology for participating in the programming has yet to be developed. Perhaps the emerging interactive systems will allow far more to participate than previously, but it is likely that the complex social structures of the future will include social hierarchies. If such is the case, then a critical question becomes, How can the elite be held accountable?

The upsurge of legislatures in recent decades does not assure that representatives will continue to be the focal persons of social structures.

Mandarins, charismatics, entertainers, and teachers compete with representatives for hegemony. The winner of that contest may be determined by the relative compatibility of the relationships with the emerging information technologies.

The computerization of bureaucracies has enhanced the ability of modern mandarins to process information and may be lessening their accountability to citizens. Most mandarins willingly serve either representatives or charismatics and use their technical skills to assure their own well-being.

For approximately two centuries in legislature-centered states, representatives and mandarins have competed with one another to define reality and program the future. Both sets of specialists may be losing their elite positions to entertainers and educators as social structures move from the national to the international arena.

State structures composed of a combination of representatives and bureaucrats are relatively new, whereas state structures based on coalitions of warriors and record keepers extend into the distant past. The endurance of state structures based on a merger of representatives and mandarins is yet to be determined.

Representative-bureaucratic state structures rest on a foundation of solidarity, they open the flow of information about state affairs, and accountability of focal persons to periphery people has merged with accountability to superordinates and the restricted flow of information. Both legislative-centered and bureaucratic structure of the past were based on written languages, but whereas computers have extended bureaucratic relationships, it is questionable that they have enhanced representative relationships.

Candidates for legislatures appeal to voters via the broadcast media and would-be-mandarins acquire access to state structures by demonstrating expertise in the management of computerized information. The compatibility of mandarins and charismatics is yet to be demonstrated. Weber presumed that they were incompatible and that successful charismatics became metamorphosized into mandarins. Such may not be the case for electronic charismatics.

Representatives and charismatics serve as catalysts; people rally around successful candidates. In confrontations, representatives and charismatics articulate, as best they can, the sentiments and interests of others. They are public figures. In contrast, despite their title of public servant, mandarins avoid appearing before the populace. Mandarins seldom stir passion in the populace. The vocabulary of representatives and charismatics is future centered. The vocabulary of bureaucrats is present centered. The bureaucratic slogan is Follow Procedures; the slogan of representatives and charismatics is A New Day Is Coming.

Elite bureaucrats supervise subordinates, attend committee meetings to rewrite regulations, draft reports to superordinates, and process information. Representatives negotiate with fellow members of legislatures, form alliances, meet with reporters, and attend functions to appease special-interest groups. Charismatics make public appearances, coalesce sentiments, define reality, and project a Utopia. During periods of quiescence the influence of mandarins expands; during crisis the influence of representatives and charismatics comes to the fore.

The increase in social distance between bureaucrats and clients associated with computer-processed information may result in even greater crises of legitimacy than those that occurred when bureaucracies processed written and printed information. One antidote to the crisis in the legitimacy of bureaucratic structures is the formation of charismatic relationships. The bureaucratic response to the cataclysmic changes proposed by charismatics is horror and fear. But once changes have been instituted, mandarins tend to accept them. Modern mandarins are as willing to form coalitions with charismatics as their forefathers were to form coalitions with warriors.

The conflicts between representatives and mandarins, in contrast to those between charismatics on the one hand and either representatives or mandarins on the other hand are seldom dramatic. Bureaucrats and representatives routinely cooperate with one another on a one-to-one basis. "Rare indeed are issues that pit politicians as a group against bureaucrats as a group" (Aberbach, Putman, and Rockman 1981:21). Nonetheless, the outcomes of these encounters in the long run usually favor the interests of mandarins.

A conflict between a head of state who has infused the state structure with a charismatic relationship and the chief mandarin of a nation is very likely to be resolved in favor of the head of state. Yet, in the long run, the interests of mandarins usually prevail. Mandarins remain in place while charismatic heads of state come and go.

The influence of the teachers is more subtle, but perhaps in the long run more significant. It is the faculty of the knowledge centers who generate knowledge that then diffuses to primary and secondary schools and the general populace. The career path of representatives and bureaucrats is through centers of knowledge. A six-nation study found 96 percent of national bureaucrats had attended a university (p. 50). Although difficult, it is possible for an elected official to acquire national stature. In a similar vein, ethnic minorities, women, and the poor have almost no chance of obtaining an elite national position in bureaucracy, but they have a least some chance of achieving success as a representative or charismatic. Many with little formal schooling achieved prominence by becoming the focal person of extended charismatic relationships.

Advancement in bureaucracies is achieved by demonstrating an ability to process referential symbols and concepts. As campaigning and governing move to the broadcast media, advancement in state structure becomes based on ability to use evocative symbols to coalesce sentiments.

Nearly all of the bureaucratic elite come to their positions through a series of incremental promotions and take hierarchical bureaucratic structures for granted. "The bureaucrat's principal reference points are upward and downward in his organization, and occasionally lateral to other sectors of the government bureaucracy" (p. 209). Politicians, in contrast, have a series of lateral relationships with fellow representatives, leaders of special interest groups, entrepreneurs, and bureaucrats. "The average member of parliament had spent roughly 70 percent of his adult life outside national government, the average top administrator less than 20 percent" (p. 69). Mavericks are exceedingly rare in bureaucracies, but on occasion are elected to legislatures and become heads of state.

Those who occupy focal positions in any social structure, whether mandarins, representatives, charismatics, or teachers, have the opportunity to manipulate what is communicated for their personal benefit. That those who occupy focal positions sometimes foist off their definitions of reality on others is not new. What is new is that would-be charismatics can use the broadcast media and effectively compete with bureaucrats who use numbers and representatives who use written narratives in offering definitions of reality. Mandarins genuflect to numbers, representatives to printed information, and charismatics to the broadcast media.

The continuation of democratic structures requires the destruction of monopolies of knowledge such as those often maintained by mandarins and the destruction of differential access to the broadcast media. The broadcast media have been used by many to create relationships that are antithetical to both mandarin- and representative-centered relationships.

The increase in the amount of information retained and circulated has allowed us to construct more complex social structures and has given us the power to do things that were unimaginable by our ancestors. That is not an unmixed blessing. Our ancestors did not have the power to eliminate themselves. On occasion they eradicated other species and societies, but the human species was not threatened by self-destruction. Today the human species could destroy the ozone layer, render the oceans dead seas, and render all land masses sterile.

PROPHECY

A few call for us to free ourselves from the chains of our technology. If that program were adopted, it would create a catastrophe greater than any previously experienced. Conversely, a few naive souls optimistically prophetize that information technologies are ushering in a harmonious global village. Perhaps the electronic media are moving us toward a global village, but villagers have never been particularly harmonious. So long as there are divisions of labor, differential wealth, and prestige rankings, disharmony will be a facet of the human condition. Some would abolish all differentiations to create a homogeneous world, but that appears to be an impossibility. If it were possible, its implementation would result in a dreary world. The instantaneous transmission of entertainment, information, and knowledge made possible by the electronic media may contribute to generating a more exciting world, but they will not resolve all the problems confronting humanity.

New social structures will have to be formed to resolve some of the problems. Negotiations and social encounters wherein people confront one another and establish mutually tolerable solutions will be necessary to resolve many of the disharmonies that pervade humanity. Information technologies can contribute to the resolution of disharmonies and conflicts, but by themselves will not resolve them.

The effective use of information technologies to resolve disharmonies requires reflectivity and reflexiveness. Information technologies are human inventions and only exist in use, but in a sense, they take on a life of their own when they become part of the taken-for-granted worlds of ordinary citizens. For example, few statisticians are self-conscious users of numeric concepts. In contrast, the Greeks of Socrates' time were self-conscious users of written language and at best a few of them reflected on its consequences. It is after a technology has become taken for granted that its format has the greatest consequences for human action and thought. The developers of and specialists in a given technology usually claim the technology opens new vistas. Each technology does provide opportunities for people to develop new forms of sociation and new experiences, yet each also constrains human association and thought. The constraints are often subtle and insignificant during the early stages of development but sometimes become blatant and significant.

From the development of numeric concepts through the development of satellite broadcasting, information technologies have emerged and matured as an adjunct to market-centered activity. Exceptions include the development of spatial concepts, the invention of written languages, and the invention of print. Spatial concepts and written languages emerged as adjuncts to temples. The significance of temple-centered activity has withered, and has been replaced by profane knowledge centers and exciting broadcast entertainment. The marketplace and state structures continue to flourish. Most significant inventions of information technologies of the past several decades have emerged from an interfacing of markets and knowledge centers.

Communities that rest on a foundation of horizontal comradeship derived from recordings and broadcastings are becoming more extensive. Temple-centered celebrations first extended comradeship beyond bands. Temple-centered celebrations have been superseded by global broadcasting systems that extend comradeship around the world as all tune in to watch soccer games.

The forms of sociation that are to pervade the lives of future generations are yet to be determined, but it is clear that they will reflect the formats offered by the electronic media. If the idea that information and communication are processes rather than static entities is kept in the forefront though, then human control of technology remains a viable alternative. In that case, change becomes an altered process, not a transformation of a resistant static entity.

CONCLUSIONS

The amount of information preserved and circulated in the past few generations has surpassed that of all prior generations. In that process perhaps we have become more human. Certainly most of us have more elaborate selves and more knowledge than any prior generation, yet there is also a greater separation of self from others than before. Some derive greater pleasure from pornographic videotapes than from sexual intercourse with a lover; some prefer the pleasures of the soap opera to a visit with a neighbor. Each technology, in other words, has been a blessing and a burden. Whichever it is has to a large extent been determined by how it has been meshed with social structures.

Several critics of the media have advanced the theme that the media constitute an environment sui genesis. They claim we attend to only artifactual visual configurations, recorded auditory sequences, and edifices such as homes, roads, automobiles, and urban complexes. The media, however, do not constitute an environment sui genesis. Even the person who spends her working hours looking at computer displays and her leisure time watching television must on occasion confront the rain and snow. Nonetheless, those who bemoan the passing of the natural environment have a point. Without an external environment, the mind becomes lost in its own subjectivity. One study found 46 percent of the children claimed they liked television more than they liked their fathers (Williams 1982). More people prefer to watch a television broadcast than spend time with friends. The human species may becoming so involved with its products that, like those in the sensory deprivation studies, it has become lost in its own creations.

Each technology has taken on a brute existence that requires people to shape their action to its format. The influence from the media is subtle, but pervasive and obdurate. The information and entertainment offered by the electronic media is more exciting and satisfying for many than that derived from the natural environment. One observes the sculpted faces on Mount Rushmore, not Mount Rushmore; one observes the video recording of a child, not the child; one watches the television broadcast of a game, not the game.

The electronic technologies may be as threatening to humanity as a loaded machine in the hands of a young child is to a family. The so-called experts have no greater understanding of the consequences of the electronic media than the first literate citizens of ancient Greece had of the consequences of the phonetic alphabet. The electronic media probably will neither destroy nor save humanity, but they will transform it. Those in the forefront of the emerging media are like a child who holds in its hand a fire extinguisher capable of saving the lives of siblings and only knows how to activate it, but not how to use it. Will we learn to use the electronic media in time?

The hope for perfect foolproof communication is a fantasy, just as were Plato's perfect forms. The only way such a fantasy could be achieved would be to divorce human life from the environment. If that becomes possible, then there will be no emergence, no chaos, no unanticipated events. That would be the end of humanity.

Perhaps the hope for the future stems from the fact that distal futures occupy a more central position in the thoughts of the elite than ever before. Whereas in the past, pleasures derived from the present dominated the thoughts of many people, many members of modern elites from Hitler to Mother Theresa are concerned with the evaluations that will be made of them by future generations.

References

Aberbach, Joel D., Robert D. Putman, and Bert A. Rockman. 1981. *Bureaucrats and Politicians in Western Democracies.* Cambridge, MA: Harvard University Press.

Adams, Robert McC. 1966. *The Evolution of Urban Society.* Chicago: Aldine.

Agrawal, Binod. 1988. "India." Pp. 83–102 in *Video World-Wide,* edited by Manuel Alvarado. Paris: United Nations Educational, Scientific, and Cultural Organization.

Alexandre, Laurien. 1988. *The Voice of America: From Detente to the Reagan Doctrine.* Norwood, NJ: Ablex.

Altheide, David L. 1985. *Media Power.* Beverly Hills, CA: Sage.

Altheide, David L. 1995. *An Ecology of Communication.* Hawthorne, NY: Aldine de Gruyter.

Altheide, David and Robert Snow. 1991. *Media Worlds in the Postjournalism Era.* Hawthorne, NY: Aldine de Gruyter.

Altschull, J. Herbert. 1984. *Agents of Power.* New York: Longman.

Alvarado, Manuel (ed.). 1988. *Video World-Wide.* Paris: United Nations Educational, Scientific, and Cultural Organization.

Anderson, Benedict. 1983. *Imagined Communities.* London: Thetford.

Aronson, Sidney H. 1977. "Bell's Electrical Toy: What's the Use?" Pp. 15–39 in *The Social Impact of the Telephone,* edited by Ithiel de Sola Pool. Cambridge, MA: MIT Press.

Attali, Jacques and Yves Stourdze. 1977. "The Birth of the Telephone and Economic Crisis: The Slow Death of the Monologue in French Society." Pp. 97–111 in *The Social Impact of the Telephone,* edited by Ithiel de Sola Pool. Cambridge, MA: MIT Press.

Aurenche, O. 1986. "Mesopotamian Architecture from the 7th to the 4th Millennia." *Sumer* 62:71–80.

Badawy, Alexander. 1965. *Ancient Egyptian Architectural Design; a Study of the Harmonic System.* Berkeley: University of California Press.

Baer, Karl. 1960. "A Note on Egyptian Units of Area in the Old Kingdom." *Journal of Near Eastern Studies* 15:113–17.

Baer, Kaus. 1960. *Rank and Title in the Old Kingdom.* Chicago: University of Chicago Press.

Bagdikian, Ben H. 1971. *The Information Machines; Their Impact on Men and the Media.* New York: Harper and Row.

253

Bailyn, Bernard. 1967. *The Ideological Origins of the American Revolution.* Cambridge, MA: Belknap Press of Harvard University Press.

Baines, John. 1982. "Literacy and Ancient Egyptian Society." *Man* (September): 572–99.

Baines, John. 1988. "Literacy, Social Organization, and the Archaeological Record: The Case of Early Egypt." Pp. 192–214 in *State and Society,* edited by John Gledkill, Barbara Bender, and Moyens Trolle Larsen. London: Unwin Hyman.

Baines, John. 1989. "Ancient Egyptian Concepts and Uses of the Past: 3rd and 2nd Millennium Evidence." Pp. 131–49 in *Who Needs the Past?* edited by Robert Layton. Boston: Unwin Hyman.

Balazs, Etienne. 1964. *Chinese Civilization and Bureaucracy.* Translated by H. M. Wright. New Haven, CT: Yale University Press.

Baldwin, John W. 1971. *The Scholastic Culture of the Middle Ages, 1000–1300.* Lexington, MA: D. C. Heath.

Bales, Robert F. 1950. *Interaction Process Analysis; a Method for the Study of Small Groups.* Cambridge, MA: Addison-Wesley.

Barnouw, Erik. 1974. *Documentary.* New York: Oxford University Press.

Batty, M. 1987. "The Spatial Impact of Computer Culture: Post-industrialism and the Spread of the Personal Computer in Britain." Pp. 185–207 in *The Spatial Impact of Technological Change,* edited by John F. Brotchie, Peter Hall and Peter W. Newton. London: Croom Helm.

Bender, Thomas. 1988b. "Afterword." Pp. 290–97 in *The University and the City,* edited by Thomas Bender. New York: Oxford University Press.

Biggs, Robert. 1974. *Inscriptions from Tell Abu Salabikh.* Chicago: University of Chicago Press.

Blatherwich, David E. S. 1987. *The International Politics of Telecommunications.* Berkeley: University of California Press.

Blumer, Herbert, 1990. *Industrialization as an Agent of Social Change: A Critical Analysis.* Edited with an Introduction by David Maines and Thomas Morrione. Hawthorne, NY: Aldine de Gruyter.

Bogert, Carroll. 1995. "Chat Rooms and Chardos." *Newsweek,* August 21:36.

Boorstin, Daniel L. 1985. *The Discoverers.* New York: Vintage.

Boston, Ray. 1988. "W. T. Snead and Democracy by Journalism." Pp. 91–106 in *Papers for the Millions,* edited by Joel H. Wiener. New York: Greenwood.

Botein, Stephen. 1980. "Printers and the American Revolution." Pp. 11–57 in *The Press and the American Revolution,* edited by Bernard Bailyn and John B. Hench. Worcester, MA: American Antiquarian Society.

Bowen, John R. 1989. "Poetic Duels and Political Change in the Goyo Highlands of Sumatra." *American Anthropologists* 91:25–40.

Boyd, Douglas A. 1989. "The Videocassette Recorder in the USSR and Soviet Bloc Countries." Pp. 252–70 in *The VCR Age,* edited by Mark R. Levy. Newbury Park, CA: Sage.

Boyd, Douglas A. and Nawaf Adwan. 1988. "The Gulf States, Jordan and Egypt." Pp. 159–80 in *Video World-Wide,* edited by Manuel Alvarado. Paris: United Nations Educational, Scientific, and Cultural Organization.

Brasch, Walter M. and Dana R. Ulloth. 1986. *The Press and the State: Sociohistorical and Contemporary Studies.* Lanham: University Press of America.

Bridson, D. G. 1971. *Prospero and Ariel.* London: Victor Gollancz.

Briggs, Asa. 1977. "The Pleasure Telephone: A Chapter in the Prehistory of the Media." Pp. 40–65 in *The Social Impact of the Telephone,* edited by Ithiel de Sola Pool. Cambridge, MA: MIT Press.

Buel, Richard, Jr. 1980. "Freedom of the Press in Revolutionary America: The Evolution of Libertarianism, 1706–1820." Pp. 59–98 in *The Press and the American Revolution,* edited by Bernard Bailyn and John B. Hench. Worcester, MA: American Antiquarian Society.

Burke, Edmund. [1790] 1968. *Reflections on the Revolution in France.* Middlesex, England: Penguin.

Burke, James. 1985. *The Day the Universe Changed.* Boston: Little, Brown.

Burks, Alice R. and Arthur W. Burks. 1988. *The First Electronic Computer.* Ann Arbor: University of Michigan Press.

Burton, David M. 1985. *The History of Mathematics.* Boston: Allyn and Bacon.

Carey, James W. 1989. *Communication as Culture.* Boston: Unwin Hyman.

Carpenter, Rhys. 1933. "The Antiquity of the Greek Alphabet." *American Journal of Archaeology* 37:8–29.

Carpenter, Rhys. 1938. "The Greek Alphabet Again." *American Journal of Archaeology* 42:58–69.

Carter, Douglass. 1959. *The Fourth Branch of Government.* Boston: Houghton-Mifflin.

Cavazza, Fabio L. 1979. "Italy: From Party Occupation to Party Partition." Pp. 76–113 in *Television and Political Life,* edited by Anthony Smith. New York: St. Martin's.

Chen, Shing-Ling. 1993. *The Self, the Community, and the Electronic Media.* Ph.D dissertation, University of Iowa, Iowa City.

Cheng, Ying-wan. 1970. *Postal Communication in China and Its Modernization, 1860–1896.* Cambridge, MA: Harvard University Press.

Cipolla, Carlo M. 1956. *Money, Prices, and Civilization in the Mediterranean World.* Princeton, NJ: Princeton University Press.

Clark, Grahame. 1986. *Symbols of Excellence.* Cambridge: Cambridge University Press.

Cmiel, Kerneth. 1990. *Democratic Eloquence.* New York: Morrow.

Collier, George A. 1982. "In the Shadow of Empire: New Directions in Mesoamerican and Andean Ethnohistory." Pp. 1–20 in *The Inca and Aztec States 1400–1800,* edited by George A. Collier, Renato I. Rosaldo, and John D. Wirth. New York: Academic.

Couch, Carl J. 1984. *Constructing Civilizations.* Greenwich, CT: JAI.

Couch, Carl J. 1986. "Markets, Temples, and Palaces." Pp. 127–59 in *Studies in Symbolic Interaction,* Vol. 7, edited by Norman K. Denzin. Greenwich, CT: JAI.

Couch, Carl J. 1987. *Researching Social Processes in the Laboratory.* Greenwich, CT: JAI.

Couch, Carl J. 1989. *Social Processes and Relationships.* Dix Hills, NY: General Hall.

Couch, Carl J. and Shing-Ling Chen. 1988. "Orality, Literacy and Social Structure."

Pp. 155–72 in *Communication and Social Structure*, edited by David R. Maines and Carl J. Couch. Springfield, IL: Charles C. Thomas.

Culley, Robert C. 1967. *Oral Formulaic Language in the Biblical Psalms*. Toronto: University of Toronto.

Czitrom, Daniel J. 1982. *Media and the American Mind—From Morse to McLuhan*. Chapel Hill: University of North Carolina Press.

Darnton, Robert. 1982. *The Literary Underground of the Old Regime*. Cambridge, MA: Harvard University Press.

Davidovits, Joseph and Margie Morris. 1988. *The Pyramids*. New York: Hippocrene.

Davidson, Ian and William Nobel. 1989. "The Archaeology of Perception." *Current Anthropology* 30 (April):125–56.

DeFrancis, John. 1989. *Visible Speech: The Diverse Oneness of Writing Systems*. Honolulu: University of Hawaii Press.

de Sola Pool, Ithiel. 1977. *The Social Impact of the Telephone*. Cambridge, MA: MIT Press.

de Sola Pool, Ithiel. 1983. *Technologies of Freedom*. Cambridge, MA: Harvard University Press.

de Tocqueville, Alexis. [1835] 1956. *Democracy in America*. Specially edited and abridged for the modern reader by Richard D. Heffner. New York: New American Library.

Diakonoff, I. M. 1983. "Some Reflections on Numerals in Sumerian: Towards a History of Mathematical Speculations." *Journal of the American Oriental Society* 103:83–93.

Dizard, Wilson P. and S. Blake Swensrud. 1987. *Gorbachev's Information Revolution*. Washington, DC: Westview.

Eisenstein, Elizabeth L. 1979a. *The Printing Press as an Agent of Change*, Vol. I. New York: Cambridge University Press.

Eisenstein, Elizabeth L. 1979b. *The Printing Press as an Agent of Change*, Vol. II. New York: Cambridge University Press.

Ellis, Peter B. 1978. *Caesar's Invasion of Britain*. London: Orbis.

Elvin, Mark. 1973. *The Pattern of the Chinese Past*. Stanford, CA: Stanford University Press.

Epstein, Edward J. 1973. *News from Nowhere: Television and the News*. New York: Random House.

Fairbank, John and Edwin Reischauer. 1989. *China: Tradition and Transformation*. Boston: Houghton-Mifflin.

Febvre, Lucien and Martin Henri-Jean. 1976. *The Coming of the Book*. Norfolk: Lowe and Brydone.

Fehl, Noah. 1964. *History and Society*. Hong Kong: Chinese University of Hong Kong.

Feigenbaum, Edward A., and Pamela McCorduck. 1983. *The Fifth Generation: Artificial Intelligence and Japan's Computer Challenge to the World*. Reading, MA: Addison-Wesley.

Feldman, Edmund Burke. 1987. *Varieties of Visual Experience*. New York: Harry N. Abrams.

Ferruolo, Stephen C. 1988. "Parsius-Paradisus: The City, Its Schools, and the Origins of the University of Paris." Pp. 22–45 in *The University and the City,* edited by Thomas Bender. New York: Oxford University Press.

Finkelstein, Jacob J. 1979. "Early Mesopotamia 2500–1000 BC." Pp. 50–110 in *Propaganda and Communication in World History,* Vol. 1, edited by Harold D. Lasswell, Daniel Lerner, and Hans Speier. Honolulu: University Press of Hawaii.

Finnegan, Ruth. 1988. *Literacy and Orality.* Oxford, England: Basil Blackwell.

Fisher, Henry George. 1989. "The Origin of Egyptian Hieroglyphs." Pp. 59–76 in *The Origins of Writing,* edited by Wayne M. Senner. Lincoln: University of Nebraska Press.

Flamm, Kenneth. 1988. *Creating the Computer.* Washington, DC: Brookings Institute.

Flegg, G. 1983. *Numbers, Their History and Meaning.* New York: Schocken.

Flugel, J. C. 1938. *The Psychology of Clothes.* New York: International Universities Press.

Foner, Eric. 1976. *Tom Paine and the Revolutionary America.* New York: Oxford University Press.

Foster, John L. 1980. "Sinuhe: The Ancient Egyptian Genre of Narrative Verse." *Journal of New Eastern Studies* (April):89–118.

Frankfort, Henri. 1948. *Kingship and the Gods.* Chicago: University of Chicago Press.

Friberg, J. 1986. "On the Big 6-Place Tables of Reciprocals and Squares from Seleucid Babylon and Uruk and Their Old Babylonian and Sumerian Predecessors." *Sumer* 42:81–87.

Gallin, Alice. 1986. *Midwives of Nazism.* Macon, GA: Mercer University Press.

Ganley, Gladys D. and Oswald H. Ganley. 1987. *Global Political Fallout: The First Decade of the VCR 1976–1985.* Cambridge, MA: Harvard University Press.

Ganley, Gladys D. and Oswald H. Ganley. 1989. *To Inform or to Control? The New Communications Networks.* Norwood, NJ: Ablex Publishing Corporation.

Gaur, Albertine. 1984. *A History of Writing.* New York: Scribner.

Gibson, Charles. 1948. *The Inca Concept of Sovereignty.* New York: Greenwood Press.

Gibson, McGuire and Robert D. Biggs. 1991. *The Organization of Power.* Chicago: The Oriental Institute of the University of Chicago.

Gillings, Richard J. 1972. *Mathematics in the Time of the Pharaohs.* Cambridge, MA: MIT Press.

Goldberg, Benjamin. 1985. *The Mirror of Man.* Charlottesville, VA: University Press of Virginia.

Goody, Jack and Ian Watt. 1968. "The Consequences of Literacy." Pp. 27–68 in *Literacy in Traditional Societies,* edited by Jack Goody. Cambridge: Cambridge University Press.

Graff, Harvey J. 1987. *The Labyrinths of Literacy.* London: The Falmer Press.

Green, M. W. 1981. "The Construction and Implementation of the Cuneiform Writing System." *Visible Language* (Autumn):345–72.

Grendler, Paul F. 1977. *The Roman Inquisition and the Venetian Press, 1540–1605.* Princeton, NJ: Princeton University Press.

Grousset, Rene. 1970. *The Empire of the Steppes: A History of Central Asia.* New Brunswick: Rutgers University Press.

Hadingham, Evan. 1979. *Secrets of the Ice Age.* New York: Walker.

Hafstaengl, Ernst F. S. 1957. *Hitler: The Missing Years, edited by Brian Connell.* London: Eyre and Spotteswoods.

Hallpike, C. R. 1979. *The Foundations of Primitive Thought.* Oxford: Clarendon.

Hammond, John H. and Austin Kill. 1987. *The Camera Lucida in Art and Science.* Bristol: Adam Hilger.

Hammond, Nicholas G. L. 1967. *A History of Greece to 332 B.C.* Oxford: Clarendon.

Harbameier, Michail. 1988. "Inventions of Writing." Pp. 253–78 in *State and Society,* edited by John Glenhill, Barbara Bender, and Mogens Trolle Larsen. London: Unwin Hyman.

Hardesty, Monica J. and Michael A. Katovich. 1986. "Two Triadic Interaction Contexts of Socialization." Pp. 269–94 in *Studies in Symbolic Interaction: The Iowa School,* edited by Carl J. Couch, Stanley L. Saxton, and Michael A. Katovich. Greenwich, CT: JAI.

Harris, Paul. 1977. *Broadcasting from the High Seas.* Edinburgh: Paul Harris.

Harris, William V. 1989. *Ancient Literacy.* Cambridge, MA: Harvard University Press.

Hart, James D. 1950. *The Popular Book: A History of America's Literary Taste.* New York: Oxford University Press.

Havelock, Eric A. 1963. *Preface to Plato.* Cambridge, MA: Harvard University Press.

Havelock, Eric A. 1971. *Prologue to Greek Literacy.* Cincinnati, OH: University of Cincinnati.

Havelock, Eric A. 1982. *The Literate Revolution in Greece and Its Cultural Consequences.* Princeton, NJ: Princeton University Press.

Havelock, Eric A. 1986. *The Muse Learns to Write.* New Haven, CT: Yale University Press.

Hayhoe, Ruth. 1989. *China's Universities and the Open Door.* London: M. E. Sharpe.

Henige, David P. 1974. *The Chronology of Oral Tradition.* Oxford: Clarendon.

Hilberg, Raul. 1961. *The Destruction of the European Jews.* New York: Harper.

Hilbert, Richard A. 1987. "Bureaucracy as Belief, Rationalization as Repair: Max Weber in a Post-Functionalist Age." *Sociological Theory* (Spring):80–86.

Hintz, Robert A., Jr. 1975. "Foundations of Social Action." Pp. 47–64 in *Constructing Social Life,* edited by Carl J. Couch and Robert A. Hintz, Jr. Champaign, IL: Stipes.

Hintz, Robert A., Jr., and Carl J. Couch. 1975. "Writing and Reading as Social Activities." Pp. 139–55 in *Constructing Social Life,* edited by Carl J. Couch and Robert A. Hintz, Jr. Champaign, IL: Stipes.

Hinz, Walter. 1973. *The Lost World of Elam.* New York: New York University Press.

Hoffman, Michael A. 1979. *Egypt before the Pharaohs.* New York: Alfred A. Knopf.

Huffman, James L. 1980. *Politics of the Meiji Press.* Honolulu: University Press of Hawaii.

Hyde, J. K. 1988. "Universities and Cities in Medieval Italy." Pp. 13–21 in *The University and the City,* edited by Thomas Bender. New York: Oxford University Press.

Ingelbart, Louis Edward. 1987. *Press Freedoms*. New York: Greenwood.

Inkeles, Alex and Raymond Bauer. 1959. *The Soviet Citizen*. Cambridge: Harvard University Press.

Innis, Harold A. 1951. *The Bias of Communication*. Toronto: University of Toronto Press.

Isfrah, G. 1985. *From One to Zero*. Translated by L. Bair. New York: Viking.

Jackall, Robert. 1987. "The Moral Ethos of Bureaucracy." Pp. 89–108 in *Bureaucracy against Democracy and Socialism*, edited by Ronald M. Glassman, William H. Swatos, Jr., and Paul L. Rosen. New York: Greenwood.

Jacobs, Jane. 1970. *The Economy of Cities*. New York: Vintage.

Jensen, Minna Skafte. 1980. *The Homeric Question and the Oral-Formulaic Theory*. Copenhagen: Museum Tusculanum Press.

Jowett, Garth, and James M. Linton. 1980. *Movies as Mass Communication*. Beverly Hills, CA: Sage.

Karttunen, Frances. 1982. "Nahuatl Literacy." Pp. 395–417 in *The Inca and Aztec States 1400–1800*, edited by George A. Collier, Renato I. Rosaldo, and John D. Wirth. New York: Academic.

Katovich, Michael A., Marion W. Weiland, and Carl J. Couch. 1981. "Access to Information and Internal Structures of Partisan Groups: Some Notes on the Iron Law of Oligarchy." *Sociological Quarterly* 2:431–46.

Katsh, M. Ethan. 1989. *The Electronic Media and the Transformation of Law*. New York: Oxford University Press.

Kees, Herman. 1961. *Ancient Egypt; a Cultural Topography*. Chicago: University of Chicago Press.

Kielbowicz, Richard B. 1989. *News in the Mail*. Westport, CT: Greenwood.

Kerferd, G. B. 1981. *The Sophistic Movement*. Cambridge: Cambridge University Press.

Ketcham, Ralph. 1986. *The Anti-Federalist Papers*. Edited and with an introduction by Ralph Ketcham. New York: New American Press.

Klopfenstein, Bruce C. 1989. "The Diffusion of the VCR in the United States." Pp. 21–39 in *The VCR Age*, edited by Mark R. Levy. Newbury Park, CA: Sage.

Kramer, Samuel Noah. 1959. *History Begins at Sumer*. Garden City, NY: Doubleday.

Kris, Ernst and Hans Speier. 1944. *German Radio Propaganda*. London: Oxford University Press.

Larsen, Mogens Trolle. 1988. "The Role of Writing and Literacy in the Development of Social and Political Power." Pp. 173–91 in *State and Society*, edited by John Gledhill, Barbara Bender, and Mogens Trolle Larsen. London: Unwin Hyman.

Larson, James F. 1984. *Television's Window on the World*. Norwood, NJ: Ablex.

Laudon, Kenneth C. 1986. *Dossier Society*. New York: Columbia University Press.

Leach, E. R. 1954. "Primitive Time-Reckoning." Pp. 110–27 in *A History of Technology*, Vol. 1, edited by Charles Singer, E. J. Homyard, and A. R. Halls. Oxford: Clarendon.

Lewis, David. 1972. *We, the Navigators*. Honolulu: University Press of Hawaii.

Lewis, David. 1978. *The Voyaging Stars: Secrets of the Pacific Navigators*. New York: W.W. Norton.

Lind, Joan. 1988. "Toward a Theory of Cultural Continuity and Change: The

Innovation, Retention, Loss and Dissemination of Information." Pp. 173–94 in *Communication and Social Structure,* edited by David R. Maines and Carl J. Couch. Springfield, IL: Charles C. Thomas.

Lind, Joan Dyste. 1989. "Information, Teams, and Social Evolution." *Cultural Dynamics* 2:79–106.

Logan, Robert K. 1986. *The Alphabet Effect.* New York: William Morrow.

Lord, Albert G. 1978. *The Singer of Tales.* New York: Atheneum.

Lumpkin, B. 1980. "The Egyptian and Pythagorean Triples." *Historia Mathematica* 7:186–87.

Mackenzie, Kenneth. 1959. *The English Parliament.* Baltimore, MD: Penguin.

MacKie, Ewan W. 1977a. *Science and Society in Prehistoric Britain.* New York: St. Martin's.

MacKie, Ewan W. 1977b. *The Megalith Builders.* Oxford: Phaidon.

Maines, David R. 1989. "Culture and Temporality." *Cultural Dynamics* 2:107–23.

Mair, Victor H. 1985. "Language and Ideology in the Sacred Edict." Pp. 325–59 in *Popular Culture in Late Imperial China,* edited by David Johnson, Andrew J. Nathan, and Evelyn S. Rawski. Berkeley: University of California Press.

Makemson, Maud W. 1941. *The Morning Star Rises.* New Haven, CT: Yale University Press.

Manning, Peter K. 1988. *Symbolic Communication.* Cambridge MA: MIT Press.

Mardin, Serif. 1979. "Modernization of Social Communication." Pp. 381–443 in *Propaganda and Communication in World History,* Vol. 1, edited by Harold D. Lasswell, Daniel Lerner, and Hans Speier. Honolulu: University Press of Hawaii.

Marlow, Eugene and Eugene Secunda. 1991. *Shifting Time and Space.* New York: Praeger.

Marshack, Alexander. 1972. *The Roots of Civilization.* New York: McGraw-Hill.

Marshack, Alexander. 1985. "A Lunar-Solar Year Calendar Stick from North America." *American Antiquity* (January):27–61.

Marshack, Alexander. 1989. "On Depiction and Language." *Current Anthropology* 30 (June):332–35.

McCain, Thomas A. and G. Ferrell Lowe. 1990. "Localism in Western European Radio: Untangling the Wireless." *Journal Communication* (Winter):86–101.

McCluskey, Stephen C. 1977. "The Astronomy of the Hopi Indians." *Journal for the History of Astronomy* (October):174–95.

McLuhan, Marshall. 1962. *The Gutenberg Galaxy.* New York: New American Library.

McLuhan, Marshall. 1964. *Understanding Media.* New York: McGraw-Hill.

McNelly, Theodore H. 1987. "'Induced Revolution': The Policy and Process of Constitutional Reform in Occupied Japan." Pp. 96–106 in *Democratizing Japan,* edited by Robert E. Ward and Sakamoto Yoshikazu. Honolulu: University of Hawaii Press.

Mead, George Herbert. 1934. *Mind, Self and Society.* Chicago: University of Chicago Press.

Mellart, James. 1967. *Catal Huyuk: A Neolithic Town in Anatolia.* London: Thomas and Hudson.

Mellart, James. 1975. *The Neolithic of the Near East.* New York: Charles Scribner's.

Mendelssohn, Kurt. 1974. *The Riddle of the Pyramids.* New York: Praeger.

Mellart, James. 1978. *The Archaeology of Ancient Turkey.* Totowa, NJ: Rouman and Littlefield.

Menninger, Karl. 1969. *Number Words and Number Symbols.* Translated by Paul Broneer. Cambridge, MA: MIT Press.

Meyrowitz, Joshua. 1985. *No Sense of Place.* New York: Oxford University Press.

Mickelson, Sig. 1989. *Whistle Stop to Sound Bite.* New York: Praeger.

Mickiewicz, Ellen. 1988. *Split Signals.* New York: Oxford University Press.

Millard, A. R. 1972. "The Practice of Writing in Ancient Israel." *Biblical Archaeologists* 470(December):98–110.

Minc, Leah D. 1988. "Scarcity and Survival: The Role of Oral Tradition in Mediating Subsistence Crises." *Journal of Anthropological Archaeology* 5:39–11.

Minow, Newton M., John Bartlow Martin, and Lee M. Mitchell. 1973. *Presidential Television.* New York: Basic Books.

Monaco, Paul. 1976. *Cinema and Society.* New York: Elsevier.

Moravec, Hans. 1988. *Mind Children.* Cambridge, MA: Harvard University Press.

Mott, Frank Luther. 1962. *American Journalism.* New York: Macmillan.

Muensterberger, Werner. 1974. *Children of the Desert.* New York: Harper & Row.

Nakayama, Shigeru. 1973. "Joseph Needham, Organic Philosopher." Pp. 23–44 in *Chinese Science,* edited by Shigeru Nakayama and Nathan Sivin. Cambridge, MA: MIT Press.

Needham, Joseph. 1969. *The Grand Titration.* Toronto: University of Toronto Press.

Nell, Victor. 1988. *Lost in a Book.* New Haven, CT: Yale University Press.

Nerone, John C. 1990. "Violence against the Press in U.S. History." *Journal of Communication* 40:6–33.

Nissen, Hans J. 1988. *The Early History of the Ancient New East.* Chicago: University of Chicago Press.

Oates, David and Joan Oates. 1976. *The Rise of Civilizations.* New York: Dutton.

O'Brien, P. J. and H. D. Christensen, 1986. "An Ancient Maya Measurement System." *American Antiquity* 51:136–51.

Ogan, Christine L. 1989. "The Effects of Technology on Communication Policy." Pp. 44–59 in *Media Use in the Information Age,* edited by Jerry L. Salvaggio and Jennings Bryant. Hillsdale, NJ: Lawrence Erlbaum Associates.

Ong, Walter J. 1982. *Orality and Literacy.* London: Methuen.

Oppenheim, A. Leo. 1977. *Ancient Mesopotamia: Portrait of a Dead Civilization.* Revised edition, completed by Erica Reiner. Chicago: University of Chicago Press.

Padgett, John F. 1980. "Bounded Rationality in Budgetary Research." *American Political Science Review* (June):354–72.

Parker, Richard A. 1950. "The Calendars of Ancient Egypt." In *The Oriental Institute of the University of Chicago Studies in Ancient Oriental Civilizations,* No. 26. Chicago: University of Chicago Press.

Parker, Richard A. 1974. "Ancient Egyptian Astronomy." Pp. 13–27 in *The Legacy of Egypt,* edited by John R. Hobson. London: Oxford University Press.

Perry, Charles R. 1977. "The British Experience 1876–1912: The Impact of the Telephone During the Years of Delay." Pp. 69–96 in *The Social Impact of the Telephone,* edited by Ithiel de Sola Pool. Cambridge, MA: MIT Press.

Pina-Cabral, Joao di. 1989. "The Valuation of Time among the Peasant Population of the Alto Minho, Northwestern Portugal." Pp. 59–69 in *Who Needs the Past?* edited by Robert Layton. Boston: Unwin Hyman.

Postman, Neil. 1985. *Amusing Ourselves to Death.* New York: Viking Penguin.

Prasad, Rai Gyan Narain. 1989. "Beginning of Agriculture: A Synchronism between Puranic and Archaeological Evidence." Pp. 150–61 in *Who Needs the Past?* edited by Robert Layton. Boston: Unwin Hyman.

Pullan, J.M. 1968. *The History of the Abacus.* London: Hutchinson.

Radway, Janice A. 1991. *Reading the Romance.* Chapel Hill: University of North Carolina Press.

Rashdall, Hastings. 1936. *The Universities of Europe in the Middle Ages.* Edited by F. M. Powicke and A. B. Emden. Glasgow: Oxford University Press.

Rawski, Evelyn S. 1979. *Education and Popular Literacy in Ch'ing China.* Ann Arbor: University of Michigan Press.

Rawski, Evelyn S. 1985. "Economic and Social Foundation of the Late Imperial China." Pp. 3–33 in *Popular Culture in Late Imperial China,* edited by David Johnson, Andrew J. Nathan, and Evelyn S. Rawski. Berkeley: University of California Press.

Rees, W. E. 1990. *Satellite Communications: The First Quarter Century of Service.* New York: Wiley.

Reese, William A. II and Carl J. Couch. 1990. "Toward a Pragmatic Theory of Numeric and Spatial Concepts." Pp. 411–39 in *Studies in Symbolic Interaction,* Vol. 11, edited by Norman K. Denzin. Greenwich, CT: JAI Press.

Reyman, Jonathan E. 1987. "Priests, Power, and Politics: Some Implications of Socioceremonial Control." Pp. 121–47 in *Astronomy and Ceremony in the Prehistoric Southwest,* edited by J. Carlson, W. J. Judge, and M. J. Young. Albuquerque, NM: Maxwell Museum of Anthropology.

Reyman, Jonathan E. 1989. "Priests, Power, and Politics: Some Implications of Socioceremonial Control." In *Astronomy and Ceremony in the Prehistoric Southwest,* edited by J. Clarson, W. J. Judge, and M. J. Young. Albuquerque, NM: Maxwell Museum of Anthropology.

Rice, Michael. 1990. *Egypt's Making.* New York: Routledege.

Rogers, Everett. 1986. *Communication Technology.* New York: Free Press.

Rudy, Willis. 1984. *The Universities of Europe, 1000–1914.* London: Associated University Presses.

Saerchinger, Cesar. 1938. *Hello America!* Boston: Houghton-Mifflin.

Sarton, George. 1952. *A Guide to the History of Science.* Waltham, MA: Chronica Botanica.

Saxton, Stanley, 1995. "The Couch Project" *Symbolic Interaction* 18:269-281.

Schama, Simon. 1987. *The Embarrassment of Riches.* New York: Alfred A. Knopf.

Schmandt-Besserat, Denise. 1981. "From Tokens to Tablets: A Re-Evaluation of the So-called 'Numerals and Writing.'" *Visible Language* 15:321–44.

Schmandt-Besserat, Denise. 1982. "How Writing Came About." Pp. 1–5 in *Zeitschrift fur Papyrologie and Epigraphik,* edited by Rudolf Habelt. Bonn: Habelt.

Schmandt-Besserat, Denise. 1986. "Tokens: Facts and Interpretation." *Visible Language* 20(Summer):150–271.

Schmandt-Besserat, Denise. 1988. "Quantification and Social Structure." Pp. 137–

54 in *Communication and Social Structure,* edited by David R. Maines and Carl J. Couch. Springfield, IL: Charles C. Thomas.

Schmandt-Besserat, Denise. 1989. "Two Predecessors of Writing: Plain and Complex Tokens." Pp. 27–42 in *The Origins of Writing,* edited by Wayne M. Senner. Lincoln: University of Nebraska Press.

Schmandt-Besserat, Denise. 1992. *Before Writing.* Austin: University of Texas Press.

Schwartz, Nancy L. 1988. *The Blue Guitar.* Chicago: University of Chicago Press.

Scribner, R. W. 1981. *For the Sake of Simple Folk.* Cambridge: Cambridge University Press.

Scribner, Sylvia and Michael Cole. 1981. *The Psychology of Literacy.* Cambridge, MA: Harvard University Press.

Seckman, Mark and Carl Couch. 1989. "Jocularity, Sarcasm, and Relationships: A Formal Approach." *Journal of Contemporary Ethnography* 18:327–44.

Sehested, Glenda J. 1975. "The Evolution of Solidarity." Pp. 99–118 in *Constructing Social Life,* edited by Carl J. Couch and Robert A. Hintz, Jr.. Champaign, IL: Stipes.

Seidenberg, A. 1963. "The Ritual Origin of Geometry." *Archive for History of Exact Sciences* 1(2):488–527.

Senner, Wayne M. 1989. "Theories and Myths on the Origins of Writing: A Historical Overview." Pp. 1–26 in *The Origins of Writing,* edited by Wayne M. Senner. Lincoln: University of Nebraska Press.

Seton, L. 1984. *The Archaeology of Mesopotamia,* rev. ed. London: Thames and Hudson.

Shlapentokh, Vladimir. 1990. *Soviet Intellectuals and Political Power: The Post-Stalin Era.* Princeton, NJ: Princeton University Press.

Simmel, Georg. 1950. *The Sociology of Georg Simmel.* Translated, edited, and with an Introduction by Kurt H. Wolff. Glencoe, IL: Free Press.

Simon, Herbert A. 1957. "A Behavior Model of Rational Choice." Pp. 241–60 in *Models of Man.* New York: Wiley.

Simpson, William Kelly. 1973. *The Literature of Ancient Egypt.* New Haven, CT: Yale University Press.

Singh, Purushottam. 1974. *Neolithic Cultures of Western Asia.* London: Seminar.

Sink, Barbara B. and Carl J. Couch. 1986. "The Construction of Interpersonal Negotiations." Pp. 149–66 in *Studies in Symbolic Interaciton: The Iowa School,* edited by Carl J. Couch, Stanley L. Saxton, and Michael A. Katovich. Greenwich, CT: JAI.

Small, Christopher. 1982. *The Printed Word.* Aberdeen: Aberdeen University Press.

Smith, Jeffery A. 1988. *Printers and Press Freedom.* New York: Praeger.

Snell, Bruno, 1982. *The Discovery of the Mind.* NY: Dover.

Sontag, Susan. 1977. *On Photography.* New York: Farrar, Straus and Giroux.

Spa, Miquel de Moraga and Daniel E. Jones. 1988. "Spain." Pp. 29–48 in *Video World-Wide,* edited by Manuel Alvarado. Paris: United Nations Educational, Scientific, and Cultural Organization.

Stearman, Allynn Machean. 1984. "The Yuqui Connection: Another Look at Siriono Deculturation." *American Anthropologist* (September):630–650.

Steinberg, S. H. 1959. *Five Hundred Years of Printing.* London: Faber and Faber.

Stoddart, Simon and James Whitley. 1988. "The Social Context of Literacy in Archaic Greece and Etruria." *Antiquity* 62:761–72.

Straubhaar, Joseph D. and Carolyn Lin. 1989. "A Quantitative Analysis for the Reasons for VCR Penetration Worldwide." Pp. 125–45 in *Media Use in the Information Age,* edited by Jerry L. Salvaggio and Jennings Bryant. Hillsdale, NJ: Lawrence Erlbaum Associates.

Strauss, Anselm. 1993. *Continual Permutations of Action.* Hawthorne, NY: Aldine de Gruyter.

Stross, Brian. 1989. "Olmec Vessel with a Crayfish Icon: An Early Rebus." Pp. 143–64 in *Word and Image in Maya Culture,* edited by William F. Hanks and Don S. Rice. Salt Lake City: University of Utah Press.

Tedlock, Barbara. 1982. *Time and the Highland Maya.* Albuquerque: University of New Mexico Press.

Thibault, Paul and Giacomo Toriano. 1988. "Italy." Pp. 49–61 in *Video World-Wide,* edited by Manuel Alvarado. Paris: United Nations Educational, Scientific, and Cultural Organization.

Thom, A. 1967. *Megalithic Sites in Britain.* Oxford: Clarendon.

Thom, A. 1971. *Megalithic Lunar Observatories.* Oxford: Clarendon.

Thom, A. and A. S. Thom. 1978. *Megalithic Remains in Britain and Brittany.* Oxford: Clarendon.

Toulmin, Stephen and June Goodfield. 1962. *Architecture of Matter.* London: Hutchins University Library.

Tsuen-Hsuin, Tsien. 1985. *Science and Civilization in China.* Cambridge: Cambridge University Press.

Turkle, Sherry. 1984. *The Second Self.* New York: Simon and Schuster.

Twitchet, Denis. 1983. *Printing and Publishing in Medieval China.* London: Wynkyn de Word Society.

Unguru, S. 1975. "On the Need to Rewrite the History of Greek Mathematics." *Archive for the History of Exact Sciences* 15(1)1:67–114.

Usami, Shoso. 1988. "Japan." Pp. 71–82 in *Video World-Wide,* edited by Manuel Alvarado. Paris: United Nations Educational, Scientific, and Cultural Organization.

Van Seters, John. 1983. *In Search of History.* New Haven, CT: Yale University Press.

Wachtel, Nathan. 1982. "The Mitimas of the Cochabamba Valley: The Colonization Policy of Huayna Capac." Pp. 199–235 in *The Inca and Aztec States 1400–1800,* edited by George A. Collier, Renato I. Rosaldo, and John D. Wirth. New York: Academic.

Waller, Douglas. 1995. "Onward Cyber Soldiers." *Time* (August 21):38–46.

Weiland, Marion W. 1975. "Forms of Social Relations." Pp. 80–98 in *Constructing Social Life,* edited by Carl J. Couch and Robert A. Hintz, Jr. Champaign, IL: Stipes.

Weiland, Marion W. and Carl J. Couch. 1986. "The Disintegration and Solidification of Newly Formed Partisan Groups." Pp. 309–21 in *Studies in Symbolic Interaciton: The Iowa School edited,* by Carl J. Couch, Stanley L. Saxton, and Michael A. Katovich. Greenwich, CT: JAI.

Weinberg, Nathan. 1990. *Computers in the Information Society.* Boulder, Co: Westview.

Weizenbaum, Joseph. 1976. *Computer Power and Human Reason: From Judgment to Calculation.* San Francisco: W. H. Freeman.

Westin, Alan F. and Michael A. Baker (eds). *Databanks in a Free Society; Computers, Record-Keeping, and Privacy.* New York: Quadrangle.

Wheatley, Paul. 1971. *The Pivot of the Four Quarters.* Chicago: Aldine.

Wheen, Francis. 1985. *Television.* London: Century.

Wiggins, James Russell. 1980. "Altersvard: The Legacy of the Press in the American Revolution." Pp. 365–72 in *The Press and the American Revolution,* edited by Bernard Bailyn and John B. Hench. Worcester, MA: American Antiquarian Society.

Williams, Frederick. 1982. *The Communications Revolution.* Beverly Hills, CA: Sage.

Wilson, Peter J. 1988. *The Domestication of the Human Species.* New Haven, CT: Yale University Press.

Wright, Arthur F. 1979. "Chinese Civilization." Pp. 220–56 in *Propaganda and Communication in World History,* Vol. 1, edited by Harold D. Lasswell, Daniel Lerner, and Hans Spieir. Honolulu: University Press of Hawaii.

Wright, Henry T. 1977. "Toward an Explanation of the Origin of the State." Pp. 49–68 in *Origins of the State: The Anthropology of Political Evolution,* edited by Ronald Cohen and Elman R. Service. Philadelphia: Institute for the Study of Human Issues.

Wright, Henry T. and Gregory A. Johnson. 1975. "Population, Exchange and Early State Formation in Southwestern Iran." *American Anthropologist* (June): 267–89.

Yabuuti, Kiyosi. 1973. "Chinese Astronomy: Development and Limiting Factors." Pp. 91–104 in *Chinese Science,* edited by Shigeru Nakayama and Nathan Sivin. Cambridge, MA: MIT Press.

Zuboff, Shoshana. 1988. *In the Age of the Smart Machine.* New York: Basic Books.

Zuidema, R. Tom. 1982. "Bureaucracy and Systematic Knowledge in Andean Civilization." Pp. 419–58 in *The Inca and Aztec States 1400–1800,* edited by George A. Collier, Renato I. Rosaldo, and John D. Wirth. New York: Academic.

Zuidema, R. Tom. 1983. "Hierarchy and Space in Incaic Social Organization." *Ethnohistory* 30:49–75.

Index